CHEMICAL VAPOR DEPOSITION POLYMERIZATION
The Growth and Properties of Parylene Thin Films

CHEMICAL VAPOR DEPOSITION POLYMERIZATION
The Growth and Properties of Parylene Thin Films

by

Jeffrey B. Fortin, Ph.D.
Rensselaer Polytechnic Institute
Department of Engineering Science
Troy, NY
and
GE Global Research Center
Niskayuna, NY

Toh-Ming Lu, Ph.D.
R. P. Baker Distinguished Professor of Physics
Rensselaer Polytechnic Institute
Department of Physics
Troy, NY

KLUWER ACADEMIC PUBLISHERS
Boston / Dordrecht / New York / London

Distributors for North, Central and South America:
Kluwer Academic Publishers
101 Philip Drive
Assinippi Park
Norwell, Massachusetts 02061 USA
Telephone (781) 871-6600
Fax (781) 871-6528
E-Mail <kluwer@wkap.com>

Distributors for all other countries:
Kluwer Academic Publishers Group
Post Office Box 322
3300 AH Dordrecht, THE NETHERLANDS
Telephone 31 78 6576 000
Fax 31 78 6576 474
E-Mail <orderdept@wkap.nl>

 Electronic Services <http://www.wkap.nl>

Library of Congress Cataloging-in-Publication

Chemical Vapor Deposition Polymerization – The Growth and Properties of Parlyne Thin
Films by Jeffrey B. Fortin, Ph.D. and Toh-Ming Lu, Ph.D.
ISBN 978-1-4419-5413-8

Contents

List of Figures

List of Tables

Preface

The interest and research in the parylene family polymers has remained strong since their invention in 1947. To date there have been a handful of reviews and encyclopedia articles covering parylene and its properties but no textbook devoted to the subject. The authors are attempting to fill the void with this textbook that provides a solid introduction to the parylene family polymers, their deposition, the deposition apparatus, and film properties.

Chapter 1 is an introduction to polymerization, parylene family polymers, the unique parylene deposition process, and the application space. Chapter 2 provides an overview of parylene deposition equipment and process control instrumentation, as well as details on an example deposition system. Chapter 3 provides the reader with a step-by-step process for depositing parylene, that, although is not meant to be applied to every system, points out the critical factors to depositing a high quality film in a well controlled manner. Chapter 4 provides details on parylene-N precursor thermodynamics and mass spectrometry fragmentation patterns for both the monomer and the dimer. Chapter 5 then goes deep into providing an understanding of the kinetics of the chemical vapor deposition process itself and presents a rate model for parylene-N deposition that can be extended to other parylene family polymers. Chapter 6 contains property data for many of the parylene family polymers as well as detailed property testing methods and results for multiple properties of thin parylene-N films. The testing methods can be used broadly for property measurement of other thin dielectric films. Chapter 7 gives an overview of non-parylene polymers that have been deposited via different forms of chemical vapor deposition.

This book is intended to be useful to both users and researchers of parylene thin films. It should be particularly useful for those setting up and characterizing their first research deposition system. It provides a good picture of the deposition process and equipment as well as information on system-to-system variations that is important to consider when designing a deposition system or making modifications to an existing one. Also included are methods to characterize a deposition system's pumping properties as well as monitor the deposition process via mass spectrometry. There are many references that will lead the reader to further information on the topic being discussed. This text should serve as a useful reference source and handbook for scientists and engineers interested in depositing high quality parylene thin films.

JEFFREY FORTIN

TOH-MING LU

Acknowledgments

We would like to thank Dr. Chung Lee and Dr. Jay Senkevich for many valuable discussions. We appreciate the contributions from many former group members including Dr. Lu You, Dr. G.-R. Yang, Dr. Y.-P. Zhao, and Dr. Peter Wu. The support of an NSF grant, IBM, and the Semiconductor Research Corporation over the years is deeply appreciated. The support of a fellowship for Dr. Fortin from Applied Materials during his study at Rensselaer is greatly appreciated. This text could not have been completed without the support of our families.

Chapter 1

INTRODUCTION

1.   Polymers and Their Applications

Polymeric materials and their applications are plentiful in the world around us. In its simplest form, a polymer molecule is one that is composed of a large number of repeating units of identical structure, termed the monomer. Bulk polymeric materials, often termed plastics, may consist of one or more types of polymer molecules and they might also include various additives, fillers, and cross-linking agents.

Polymers have been used in many applications in industries such as the textile, automotive, aerospace, marine, and electronics industries. New applications utilizing the unique properties of polymers are constantly emerging. Recent applications include those in the growing biotechnology, microelectronics, and photonics industries. In the biotechnology industry, polymers are being used in applications such as separation membranes, inert coatings, artificial organs, biosensing, cell biology, tissue engineering, and in drug delivery. In the microelectronics industry polymers are being used in applications such as PC-board materials, dielectrics, chemical resistant coatings, electronics displays, flexible circuits, and in micro-electro-mechanical systems (MEMS), to name a few. In the photonics industry polymers are finding applications such as wave-guide materials, photonic band gap structures, as well as in packaging.

Many of the applications for polymers today require them to be in the form of thin films, films anywhere from a few angstroms to tens of microns in thickness. The methods of growing or depositing thin polymer films have to be able to produce the desired film thickness and uniformity over the desired coating area and, at the same time, be economically feasible. The polymer material must be chosen or developed based on the requirements of the specific application. Requirements may include stringent electrical, mechanical, chemical, and optical properties.

Poly-para-xylylene based polymers, also known as parylenes, are polymers that can be deposited in thin film form at room temperature by chemical vapor deposition and are capable of providing unique properties. They have found may applications since their invention in the late 1940's and are continuing to find more applications today, particularly in the biotechnology and microelectronics industries.

2.   Polymerization Mechanisms

Polymers are often classified based on the polymerization kinetics used to produce the polymer. According to this scheme, all polymerization mechanisms are classified as either step growth or chain growth. A step growth polymerization is one that is defined to have a random reaction of two molecules that may be any combination of a monomer, an oligomer (polymer chain with less than 10 units), or a long-chain molecule. A chain growth polymerization is one that is defined to have a polymer chain that grows only one unit at a time, by the attachment of a monomer to a chain end. The chain end could be a radical, cation, or anion. Chain growth polymerization takes place in three common steps- initiation, propagation, and termination.

Parylene polymerization is of the chain growth type except that the chains are not terminated during growth. Un-reacted chain ends are buried in the film as it grows. Subsequent termination of the radical chain ends can occur post-deposition via reactions such as with atmospheric oxygen that has diffused into the polymer film [1].

3.   Polymerization Techniques

The principal polymerization techniques are bulk, solution, suspension, and emulsion polymerization [2, 3, 4]. In bulk polymerization only a liquid monomer and a monomer soluble initiator are used in the polymerization process. In solution polymerization the monomer and initiator are dissolved in an appropriate solvent, or water, which facilitates the removal of heat evolved during the polymerization process. In suspension polymerization, the monomer and initiator are suspended rather than dissolved in an inert liquid in which they are insoluble, often water, and this provides increased heat dissipation. In emulsion polymerization droplets of monomer are dispersed in water with the aid of a surfactant and then initiated using a water-soluble initiator. Chain-transfer agents are often used in all of these polymerization techniques to control the molecular weight.

In addition to these usual methods of solution polymerization, some polymers can be prepared from the gas or vapor phase with or without the assistance of a plasma. Examples of polymers produced by the vapor phase technique without a plasma are the parylenes, Teflon AF, poly(naphthalene), poly(fluorinated naphthalene), poly(octafluoro-bis-benzocyclobutane), and polyethylene [4, 5, 6]. This method of polymerization has been termed vapor phase deposition, gas phase deposition, and chemical vapor deposition (CVD).

In the CVD of a polymer, typically, a vapor of reactive monomer is produced in some fashion and then subsequently or simultaneously introduced into the deposition system where it will adsorb onto the substrate and polymerize. There is typically no by-product of the polymerization. Among the above mentioned CVD deposited polymers only parylene has been well studied and has found commercial uses. Throughout the remainder of this book the parylene deposition process will be referred to as a CVD process.

Many polymers can also be obtained via plasma polymerization or plasma CVD. In this process thin films or powder polymers are produced or deposited on a surface by contacting this surface with a glow discharge of organic or organometallic monomer. Many different starting monomers may be used including hydrocarbons with and without polar groups, fluorocarbons, those containing silicon, and those containing metal [8]. It is important to note that the term monomer is often used here but not always appropriate because there is typically not a full retention of its structure or a repeat unit in the deposited films. Plasma polymerized polymers are also finding uses in industries such as biotechnology, microelectronics, and photonics [8, 9].

The chemical vapor deposition of polymer thin films, particularly that of parylene, has some clear advantages over the other polymerization techniques in terms of coating a substrate. In bulk, solution, suspension, and emulsion polymerization coatings of the polymer are applied to a substrate via methods such as dipping, spraying, or spin-on. In these methods, a process step such as heating must be used to remove the solvent or initiate the polymerization. Problems with these processes can be non-uniform film thickness, pinholes, and non-conformality of the polymeric film to the substrate, as well as residual solvent left behind. In the CVD of parylene there is no solvent used or by-product produced. The deposited films are conformal and the monomer molecules can penetrate deep into crevices to form pinhole free films. Parylene has been shown to fill deep sub-micron gaps and produce uniform coatings over large substrate areas[5, 7].

4.   Parylene Family Polymers

4.1   Parylene History and the Gorham Method

The deposition of parylene onto a surface using a gaseous precursor was first observed by Szwarc in 1947 when he found the polymer as one of the products formed in the vacuum thermal decomposition (pyrolysis) of para-xylene, a common solvent [10, 11, 12]. Szwarc postulated that the species produced by the decomposition in the vapor phase responsible for forming the polymer was para-xylylene, and proved it to be so by mixing the deposition vapors with iodine vapor and finding para-xylylene di-iodide as the only product [10, 11]. The yields of polymer film were only a few percent even at relatively high pyrolysis temperatures, ranging from 700°C to 900°C [12].

Gorham later found a much more efficient route to the deposition of parylene films through the vacuum pyrolysis of di-para-xylylene, also termed the dimer and otherwise known as [2,2]paracyclophane [13]. He found that at temperatures above 550°C and at pressures less than 1 Torr, the dimer is quantitatively cleaved into two para-xylylene monomer units which are adsorbed onto a surface at room temperature and spontaneously polymerize yielding high molecular weight, linear parylene thin films, see figure 1.1.

4.2   Types of Parylene and Applications

The monomer for poly-para-xylylene (also termed parylene-N) is composed of an aromatic group with methylene groups attached at the para positions as shown in figure 1.1. There are many other parylene based monomer types which have the same basic structure but have replaced the aromatic or aliphatic hydrogen atoms with other atoms or chemical groups. For example there is parylene-C which has one chlorine on the aromatic and parylene-D which has two. Figure 1.2 shows a number of different monomers that have been studied to date but certainly not all [13, 14]. Out of all the monomers that have been studied those receiving significant attention are parylene-N, parylene-C, parylene-D, and the fluorinated versions [17, 15, 16, 18, 19]. The types that have been and are being used most by industry are parylene-N and parylene-C. Parylene-F (or AF-4, for aliphatic fluorine) is a version of parylene with the aliphatic hydrogens replaced with fluorine and is known for its high temperature stability in air as well as nitrogen and vacuum. These common designations initially referred to polymers deposited using Union

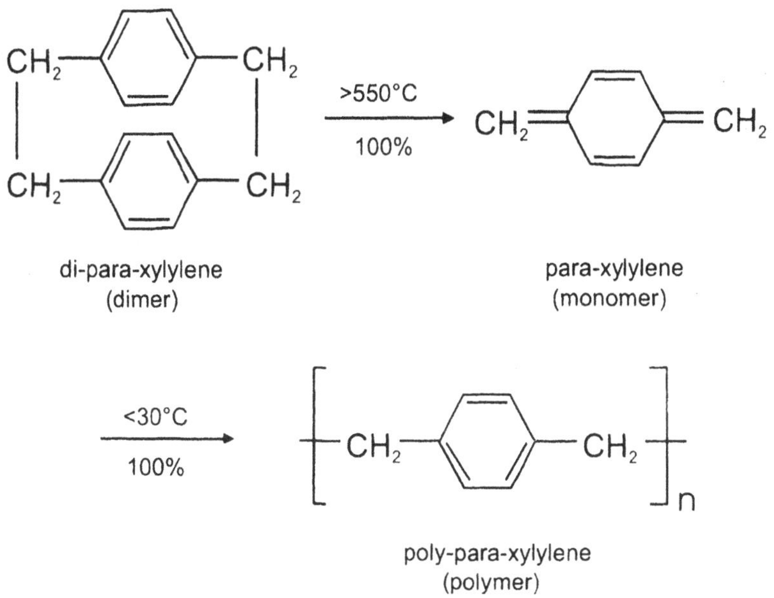

Figure 1.1: The polymerization route for parylene-N.

Carbide dimers but are now used more generally. A more recently developed fluorinated version termed parylene-VT4 has ring fluorination instead of aliphatic[169].

Recent reports in the literature show techniques for the synthesis of functionalized [2,2]paracyclophane and the subsequent sublimation and pyrolysis of these compounds to deposit a number of novel films via CVD. In one study thirteen functionalized parylene polymers were reported. The intent of the work was to produce polymers that could be deposited via CVD that could then be functionlized for biomedical applications [68, 69, 70, 71]. Another recent article reported the deposition and properties of parylene-E, a polymer film with the approximate composition of 69% diethylated and 25% monoethylated poly(p-xylylene)[67].

Co-polymerization of parylene has been demonstrated both in solution polymerization and in CVD. Errede and Hoyt prepared a co-polymer via solution polymerization of parylene and the vinyl polymer polystyrene as well as copolymers of parylene and maleic anhydride, diethyl fumarate, maleate, and acrylonitrile [20]. Gorham synthesized copolymers composed of two parylene family polymers by pyrolyzing the respective dimers and leading them to the deposition chamber [21]. Gaynor *et al.* have co-deposited parylene-C with the vinyl polymer poly(N- phenylmaleimide) and with poly(perfluorooctylmethacrylate) [22, 23]. Also, Taylor *et al.* have co-polymerized parylene-N with vinylic monomers in

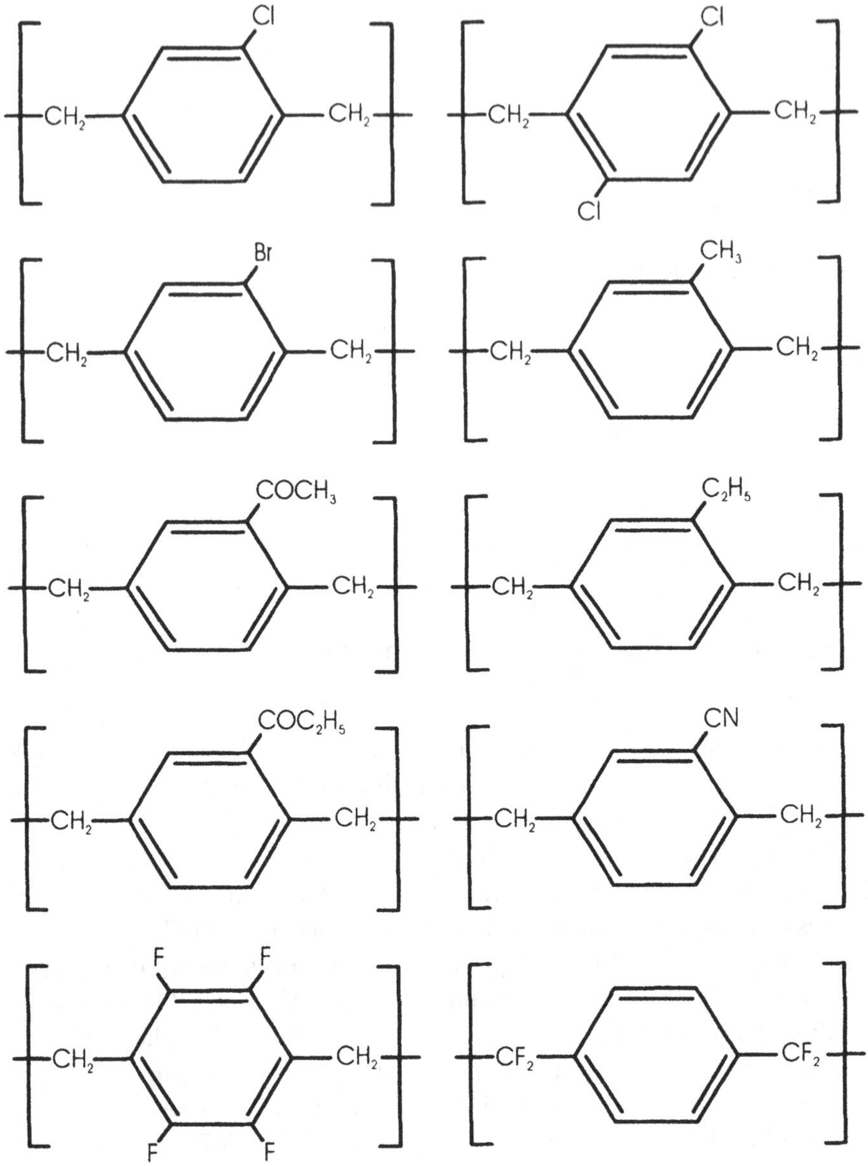

Figure 1.2: The structure of several different parylene repeat units.

an attempt to make a parylene co-polymer with a dielectric constant lower than that of parylene-N [24]. It is evident from this work that parylene monomer can be used as an initiator for gas phase deposition of other volatile monomers that would not spontaneously polymerize.

This could potentially allow for a large number of polymers to be deposited by CVD.

Applications of parylene films have been plentiful. The Gorham process produces chemically resistant, conformal, pinhole free films without the use of a solvent. These films have good mechanical and electrical properties which, for the more common types of -N,-C, and -D, have been investigated in detail. As was shown in the recent research discussed above many types of functionalized parylenes can be produced that have tailored properties. Parylene films have been used in a variety of applications including moisture barriers, chemical and corrosion resistant coatings, dielectrics for capacitors, electrical insulators for rotors and stators, electrets, coatings for artifact conservation, contamination control, and even as dry lubricants [1, 25, 26]. Parylene is also used and has been researched for several space applications [27, 28].

Parylene films have been investigated for many new applications in the last few years. Some examples are as follows: as a chemically resistant coating for piezoelectric and micromachined Si-based sensors [29, 30], as a coating on micro-machined membrane particle filters to improve their mechanical strength [31], as the bellows material in a bellows valve [32], as part of a optical modulator utilizing parylene's electrical properties [33], as part of optical and spectral filters utilizing parylene's optical properties [34], as a low dielectric constant interlayer dielectric for advanced VLSI interconnect technology [35, 36, 37, 38], as an anti-stiction layer for micro-electro-mechanical systems (MEMS) utilizing it's low coefficient of friction [39], as a material in microfluidic structures utilizing its unique deposition technique and chemical properties [170, 171], and as a bio-compatible coating [40, 175, 178, 43].

The future of the parylene family of polymers is wide open. The conformal nature of the CVD polymer films can be very useful and has many advantages over other deposition techniques. Techniques have been shown that can be used to modify or functionalize the starting dimer to tune the chemistry of the deposited films and allows for some degree of control over it's properties. Thus a number of polymers with varying chemical, mechanical, electrical, and optical properties can be deposited by CVD and that opens up the door to numerous new applications that were previously not possible.

Chapter 2

DEPOSITION EQUIPMENT

1. Deposition System Design

The main pieces of a parylene CVD system are the sublimation furnace, the pyrolysis furnace, the deposition chamber and the vacuum pumping system. There are other necessities as well, like a deposition chamber, pressure measurement system and heating tapes to keep areas of the system from having dimer or polymer deposited on them. Of course one also wants easy access to load the dimer, as well as a good design for a substrate holder that will allow for a variety of substrates to be coated with the best uniformity. An experimental system can easily be designed and put together in the lab. There are a handful of suppliers of quality vacuum system components and furnaces that one can choose from at reasonable prices. There are a few good suppliers of used vacuum equipment as well if the budget is tight. There is a great text out on vacuum technology that all users of vacuum systems should read. It is titled *"A User's Guide to Vaccum Technology"* by John F. O'Hanlon [101].

In most parylene systems there is an easy access port, some form of a flange with a butterfly style nut, at the end of the sublimation chamber where dimer can be placed in a stainless steel boat. The diameter of the sublimation tube is typically around 1.5 inches or so depending on the size of the system and most of the time it is constructed of stainless steel. The sublimation furnace needs to be capable of heating the dimer up to a maximum of about 200°C. The sublimation tube can run directly out of the sublimation furnace and into the pyrolysis furnace or there can be a connection. It is recommended that if there is a connection here that it be one with a metal gasket because of the temperatures that it will be exposed to. If there is a gap between the sublimation furnace and the pyrolysis furnace it is recommended that this area be heated with heat tape to make sure it is above the sublimation temperature to keep the amount of deposited dimer to a minimum.

Some commercial systems have a high temperature valve between the sublimation and pyrolysis furnaces. Although typically they don't use this valve to provide variable flow control but rather it is either open or closed. The authors recommend that extra money be spend in a home built system to install a high temperature heated valve at this position that can be either manually or automatically positioned to provide control over the flow rate of dimer into the system and hence control the final deposition pressure. Most deposition systems on the market do not allow for good step functions in the deposition pressure and this can lead to poor control over film thickness and poor run-to-run repeatability.

The pyrolysis furnace for most systems is in the 24 to 40 inch length range and must be capable of heating the pyrolysis zone up to temperatures of a maximum of 700°C. When depositing chlorinated or fluorinated version of the polymer it is a good practice to line the deposition tube with quartz to prevent any corrosion of the stainless steel. It is also important to calibrate your pyrolysis tube temperature profile by inserting a thermocouple or thermocouples into the tube and recording the actual temperature versus the set point as a function of position in the tube. This should also be performed for the sublimation furnace.

The deposition chamber should be easy access, like a bell jar type, or any other design that allows for quick sample transfer. Parylene will deposit everyone in this chamber unless the surface is heated to above the deposition ceiling temperature for the given material and deposition parameters. Remember that the flow of dimer into the deposition chamber and then out into the vacuum pump/cold trap will create a non-homogenous pressure distribution in the chamber. In most cases, because of the chamber geometries and the pressure range, this is not easily modeled. The system that the authors have personally seen with the best uniformity on a 6 inch silicon substrate was designed with a shower head that allowed the monomer flow verticaly down onto a wafer and the outlet was directly below the wafer chuck in line with the shower head. Depending on your application thickness uniformity on your substrate may or may not be a critical factor for you. A diffusor, which is basically a metal plate, can be installed in the deposition chamber in front of the pryolysis tube to change the flow out of the tube and create a better more uniform pressure in the chamber. It is best to experiment to find the best setup for your application.

2. Vacuum Systems

The basic vacuum system for parylene CVD consists of a mechanical pump and a cold trap. The mechanical pump can pump the system into the low mtorr range and the size required depends on the volume of the chamber and expected operating pressure and flow rates. Typical operating pressures are in the mTorr range with pump speeds in the 1 to 50 Liters/sec range. The cold trap is used to collect monomer that doesn't get deposited in the deposition chamber. This should be cleaned routinely for the best efficiency. To get to lower base pressures one can use a diffusion pump backed by a roughing pump or, alternatively, a turbo pumped system. Both of these types of systems have been successfully used on parylene deposition systems to reach lower base pressures and keep the deposition chamber to lower levels of oxygen, which can get incorporated into the growing film if it is present.

3. Pressure Measurement

Pressure measurement is needed to accurately monitor the monomer pressure in the deposition chamber. The deposition pressure is directly related to the deposition rate, as is discussed in the Kinetics chapter. Pressure may be measured by direct or indirect measurements. Direct pressure measurement is the most accurate for parylene, see "System-to-System Variations" section below. Direct pressure measurement may be accomplished via a heated capacitance monometer, which can be rather expensive. It is however imperative that it be heated otherwise parylene will deposit on the diaphragm and it will drift out of calibration. Indirect measurement can be accomplished by a thermocouple gauge or a Pirani guage (both of these have heated elements).

4. Pump Speed, Flow Rate, and Contact Time

This section will give a method that can be used to measure the pumping speed and flow rate of the dimer and monomer through a deposition system. Once the flow rates are determined an estimate of the contact time of the dimer in the pyrolysis tube can be made over a range of deposition pressures. Data is presented for the Example Deposition System (see figure 2.6) with known parameters and 100 percent conversion as a reference point. This type of exercise can be important for calibrating a new deposition system or process to insure you have 100 percent dimer-to-monomer conversion and at the same time minimizing the thermal budget.

4.1   Determining the pumping speed and flow rate

The flow rate of a gas through a tube can often be measured directly using a flow meter in line with the flow. In the case of parylene deposition measuring the flow rate of the dimer out of the sublimation tube would require a flow meter heated to around 200°C. These devices cost nearly $10,000 so this approach is not always economically feasible in many cases. Another option is to put a meter between the pyrolysis tube and the chamber but this would also require heating to at least 100°C and a large diameter orifice. Instead of measuring the flow directly with a meter the flow can determined indirectly using methods detailed by O'hanlon [101].

The approximate pump speed can be measured at the deposition chamber by using the following method:

1 The system is pumped to its base pressure, P_1, the vacuum valve is closed, and the pressure rise dP_1/dt is measured.

2 The vacuum valve is opened, the system is pumped backed to base pressure, and gas is admitted until a stable pressure, P_2, is reached.

3 The vacuum valve is closed again and the pressure dP_2/dt is measured.

4 The system volume is estimated or measured.

After these steps have been followed the pumping speed (S) can be calculated by using the equation

$$S = V \left(\frac{(\frac{dP_2}{dt} - \frac{dP_1}{dt})}{(P_2 - P_1)} \right) \tag{2.1}$$

in which S is pump speed in L/sec, V is the volume in L, P is pressure in mtorr, and t is time in seconds [101].

Once the pump speed has been determined the flow rate, Q, at a given pressure, P, can be calculated from $Q = SP$. In this equation Q is in units of L-Torr/sec and can be converted to sccm, which is standard cubic centimeters per minute or 1 cm^3 of gas at 1 atm and 25°C [101].

The above procedures were followed for the example deposition system (see figure 2.6). The pressure rises were measured and automatically recorded via computer for a number of pressures in the typical deposition range. The nearly instantaneous pressure changes as a function of time were found from the slope of the pressure versus time data over a period of a few seconds. The volume of the system was measured by attaching a small reservoir of known volume to the evacuated chamber with a valve

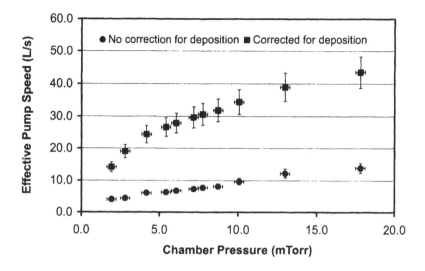

Figure 2.1: The effective pumping speed versus chamber pressure.

in between. This valve was opened to allow a known volume of air to enter the chamber and the pressure rise was recorded. The volume was then calculated using gas laws and found to be 117 L or about 30 gallons.

There were some non-trivial aspects to this pump speed measurement technique due to the nature of the gas (monomer). The monomer is continuously depositing on the walls of the chamber. This means that when measuring dP_2/dt although the pressure is increasing in the chamber as more monomer flows in the pressure is also decreasing as monomer deposits on the walls. In order to correct for this a third pressure measurement was taken. This was performed by opening valve A enough to get the pressure in the chamber up to 20 mtorr and the closing both valve A and the high vacuum valve. The pressure drop in the chamber was then recorded and dP_3/dt values were calculated from the slopes found for a period of a few seconds. Next dP_2/dt in equation 2.1 was replaced by $dP_2/dt - dP_3/dt$ (note the minus sign is used because dP_3/dt was negative). This essentially adds the pressure loss due to deposition to the measured pressure rise, dP_2/dt.

The results are shown in figures 2.1 and 2.2. Included in the figure is data for both correcting and not correcting for the deposition of monomer on the chamber walls. The pump speed is referred to as the Effective Pump Speed because part of the pumping action is really due to the deposition of the monomer on the walls and not strictly from the pumping system itself.

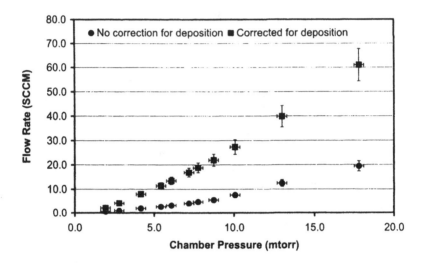

Figure 2.2: The flow rate versus chamber pressure.

4.2   Calculating the residence time

The residence time of a gas in a deposition chamber or in a tube through which the gas flows into the chamber can be calculated using the expression $t = L/V_m$. In this equation, t is the contact time in seconds, L is the length o. the tube in cm, and V_m is the rate of gas flow in mL/cm^{2}sec and can be found from $V_m = D/S$ where D is the total gas flow rate in ml/sec and S is the cross sectional area of the tube in cm^2 [90]. This calculation was made for the example deposition system in order to find the dimer contact time in the pyrolysis furnace. The length of the furnace tube is 81.3 cm and the cross sectional area is 8.96 cm^2. The flow rate, D, used was that calculated above. The results are shown in figure 2.3. This contact time is pertinent in trying to insure 100% dimer to monomer conversion. This is due to the fact that the conversion percentage is affected by contact time. For example, for low percent conversions if the contact time were to increase the conversion percent would increase. This information and data for the example deposition system should be helpful to researchers or process development engineers trying to determine the appropriate temperature range for achieving 100% conversion without an excessive thermal bud-

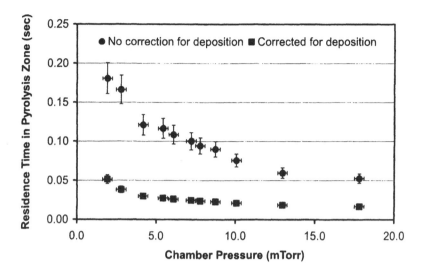

Figure 2.3: The calculated contact time of the dimer in the pyrolysis zone.

get (see section on dimer conversion). Excessive temperatures can also provide enough energy to break more bonds than just those for dimer to monomer conversion, leaving your monomer degraded.

5. Mass Spectrometry for in-situ Chemical Analysis and Process Monitoring/Control

Differentially pumped quadrupole mass spectrometers have been used for process analysis in many chemical vapor deposition (CVD) processes, including, for example, plasma enhanced CVD of silicon and silicon dioxide [74, 75, 76]. A quadrupole mass spectrometer (QMS) has a maximum operation pressure of around 10^{-4} Torr while the pressure in the deposition chamber during a typical poly-para-xylylene deposition is in the low mTorr range. Therefore, in order to utilize a QMS to analyze the deposition chemistry, it is necessary to differentially pump the quadrupole mass spectrometer, as was done on the example deposition system, see figure 2.6. A QMS is a very useful tool for in-situ chemical analysis and monitoring or control of a CVD process.

A typical mass spectrum of a single gas or a gas mixture for a QMS utilizing electron impact ionization will contain peaks due to parent ions as well as those due to fragmentation (and possibly rearrangement) of the parent ions [86]. Fragmentation can occur when the molecule absorbs energy from the ionizing electron that is in excess of the molecule's ionization energy. The degree of fragmentation for a given molecule

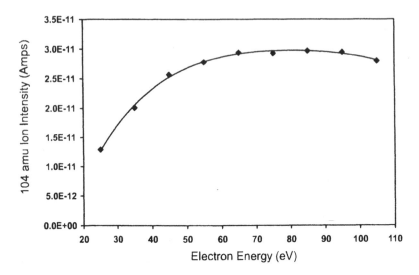

Figure 2.4: The ion intensity of the 104 amu peak (parylene monomer) versus electron energy.

can be influenced by the ionizing electron energy, the electron emission current, the focus voltage, the instrument structure, and the operating temperature [87, 88].

As the ionization electron energy is increased the fragmentation as well as the ionization probability increase. An example of this is shown in figure 2.4. In this figure, the ion current of the 104 amu monomer peak is shown as a function of electron energy. As the electron energy is increased, the amplitude of the peak increases because the probability of ionization increases. The amplitude then decreases due to increased fragmentation and possibly double ionization.

6.   Substrate Holders and Temperature Control

Any clean, non-outgasing platform can be used as a substrate holder. The holder will get coated with parylene during each run and periodic cleaning is necessary. The deposition rate for parylene is a strong function of temperature (to be discussed further in the Kinetics Chapter), therefore it may be advantageous to control the temperature of the substrate. If the substrate is flat like a silicon wafer a nice cold chuck can be designed and made using thermoelectric coolers or one can be purchased from a tooling company. If the substrate is not flat it becomes much more difficult to cool it homogenously.

7.   System-to-System Variations

Understanding system-to-system variations is important if one is trying to compare their deposition data to those from other systems. It is also important to understand how a chamber modification can affect the deposition process. Calibration runs should always be performed after any design changes have been implemented into a deposition system.

The designs of different parylene deposition systems can vary greatly from bell-jar type deposition chambers to tubes in-line with the pyrolysis furnace. In all of these systems pressure gradients exist, as they are the driving force that causes dimer to flow from the sublimation chamber into the deposition chamber. It is well known that achieving a uniform parylene film, even over a small diameter wafer for example, can be problematic. Film uniformity for a 4" wafer, for example, can run from as low as a few percent to as high as 50% depending on the deposition conditions. The reason for the non-uniform film is most likely pressure variations within the deposition region, or temperature variation for deposition taking place on a cooled substrate.

Modelling of each deposition system is theoretically possible but it can be difficult for pressures typical of parylene depositions (low mtorr). This is because in this pressure region the system is probably operating in the intermediate flow region between molecular-flow and continuum flow. In this region calculations of pressure gradients become troublesome and are not well developed [101].

Not only can the pressure vary within the chamber in an unknown manner but the methods of measuring that pressure can produce varied results. Devices like a capacitance monometer measure the pressure directly while devices like a thermal conductivity gauge measure the pressure indirectly, by measuring the thermal properties of the gas. The indirect measurement gauges need to be correctly calibrated for each gas that they measure and if this is not performed significant errors can result.

The parylene systems that are available on the market typically use thermal conductivity gauges to measure pressures but they are not corrected for the deviations from measuring nitrogen to measuring the monomer. The data that has been presented to date in most literature (including those discussion deposition kinetic rate models) was gathered using this type of gauge probably without correction (corrections of pressure measurement have not been often stated). The example deposition system described in the next section was equipped with a heated capacitance monometer. This monometer has been used to calibrate a thermocouple style indirect reading gauge. The results are shown in figure 2.5. In this figure the values of pressure measured simultaneously

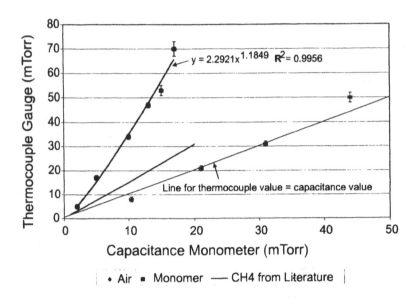

Figure 2.5: Pressure measured by a capacitance monometer
           and a thermocouple gauge for air and parylene-N
           monomer.

using a thermocouple style gauge and the capacitance monometer are
shown for both air and the monomer. For air, which is mostly nitrogen, the gauges should measure the same value while they should differ
for the monomer. Also included is correction data for CH_4 given by
the manufacturer of the thermocouple gauge for reference. There was
a large difference between the measured values of pressure between the
two gauges for the monomer. This shows that it is very important to calibrate the pressure gauge, especially when using the data for modelling
purposes.

The process of correctly measuring the pressure in the deposition
chamber at the substrate location is a difficult one. Deviations from
the true values can be large and often undeterminable. This needs to be
taken into account when modifying or upgrading a system, as mentioned
earlier.

8.   Example Deposition System

One of the vacuum deposition systems used for the authors' research
work is shown in figure 2.6. This system is home-made and consists
of a sublimation furnace, a pyrolysis furnace, a bell-jar type deposition
chamber, and a high vacuum pumping system, as well as a residual gas
analyzer and a temperature controlled substrate holder. The sublima-

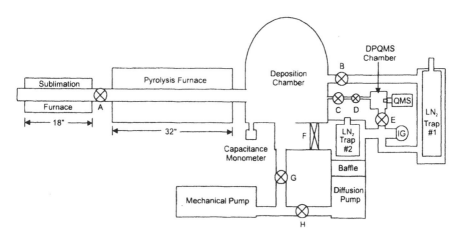

Figure 2.6: The Example Parylene Deposition System.

tion furnace is a 18 inch long Thermolyne 21100 tube furnace and the pyrolysis furnace is a 32 inch long Lindberg tube furnace. Both furnaces are controlled via digital controls and were calibrated in their respective operating ranges. A stainless steel 1.3" inner diameter tube passes through the heating zone of each furnace and connects to the bell-jar chamber. The tube is connected to the bell-jar using a conflat-style flange with a copper gasket. The volume of the bell jar was determined to be 117 ± 1 L by measuring the change in pressure when adding a known volume of air to the evacuated chamber.

A quick-flange style connection with a viton gasket allows dimer to be loaded into the sublimation furnace easily. A high temperature adjustable-flow valve is used between the sublimation furnace and the pyrolysis furnace to control dimer flow out of the sublimation furnace and therefore control the chamber pressure. This form of pressure control is not present on a typical industrial parylene deposition systems which usually adjust the dimer temperature to control the chamber pressure. The valve used on this system allows for quicker response and good control of the deposition pressure. This valve and all other tube surfaces between the sublimation and pyrolysis furnaces are heated to $200^{\circ}C$ using heater tape. This keeps the dimer from depositing on these areas. The tube is also heated to $200^{\circ}C$ from where it exits the pyrolysis furnace up to where it connects with the chamber. The temperature of all surfaces heated with heater strips is monitored using a multichannel thermocouple meter.

An aluminum shield is situated inside the chamber in front of where the monomer exits the pyrolysis tube. This shield acts as a diffusor to spread out the monomer as it flows out of the pyrolysis zone.

The system is pumped with a 4" Veeco diffusion pump backed by a Welch 1397 mechanical pump. There is a large (10 L) liquid nitrogen cold trap (#1) that is used to trap monomer that passes through the chamber without deposition. It also traps water vapor and helps to lower the base pressure of the system. There is also a smaller liquid nitrogen cold trap (#2) that is part of the diffusion pump and is used to condense any back-streaming diffusion pump oil. The diffusion pump uses Fomblin perfluoropolyether pump fluid. The base pressure in the deposition chamber is in the mid 10^{-7} Torr when high vacuum valve F is open and valve B is closed and in the high 10^{-6} Torr when high vacuum valve F is closed and valve B is open and the liquid nitrogen traps are filled.

The pressure in the deposition chamber was measured using a capacitance monometer, a direct pressure measurement device. The monometer used is a MKS Systems Inc. Baratron type 628B, which is heated to 100°C. This is a high accuracy monometer that allows pressure to be measured from 0.01 to 1000 mTorr. The heated versions are used to keep the polymer from depositing on the monometer's surface.

The monometer is connected to the deposition chamber at the base of the bell jar. It's output voltage is measured using a National Instruments data card installed in a computer. The voltage is converted to a pressure and the data can be displayed and saved to file. This monometer was calibrated at the factory and could be manually zeroed. It is a good practice to have the monometer recalibrated by the factory every few years.

The pressure in the deposition chamber during a poly-para-xylylene deposition is typically in the low mTorr range. A quadrupole mass spectrometer has a maximum operation pressure of around 10^{-4} Torr. In order to utilize a quadrupole mass spectrometer (QMS) to analyze the deposition chemistry it was therefore necessary to differentially pump the QMS. The differentially pumped quadrupole mass spectrometer (DPQMS) used here is attached to the deposition chamber via bellows sealed valve C with an orifice of 4.4 mm in line with metering valve D with an orifice of 0.79 mm when fully open. The DPQMS chamber is pumped with the diffusion pump and is typically in the high 10^{-7} to low 10^{-6} Torr range.

The QMS was purchased from Stanford Research Systems (type RGA 300) and is capable of detecting molecules with a mass number of up to 300 atomic mass units (amu). It has less than 1 amu resolution at 10% peak height per AVS standard 2.3. It is operated via a real-time Windows based software package that is used for data acquisition and analysis. A dual thoriated-iridium filament is used for electron emission and a Faraday cup sensor is used for measuring ion currents. For typical experimentation, the ion energy is set to 12 eV, the focus voltage is set to 90 V, the emission current is set to 1.0 mA, and the electron energy is set to 70 eV. All of these parameters can be changed via software.

During use, the gases present in the deposition chamber are pumped into the DPQMS chamber by opening valve C fully and adjusting metering valve D to the desired setting. The mass spectrometer system is heated via heating strips to 150°C from where it connects to the deposition chamber to valve E.

The main substrates holder consists of a large 8 inch square copper plate that is elevated from the floor of the chamber using 10 inch long stainless steel rods. This holder was used for room temperature depositions. Another substrate holder is designed for cooling and heating a silicon (or any wafer) substrate. This temperature controlled holder was made by sandwiching a thermoelectric cooler module purchased from Melcor (CP series) between two thin polished copper 6 inch square plates. The surface of the top plate was lapped and polished. A 4 inch wafer is held in intimate contact to the top copper plate by using a thin brass cover plate containing a circular aperture with an inner diameter just slightly less than 4 inches. This cover plate is positioned over the 4 inch wafer and firmly connected to the top copper plate using stainless steel bolts. The bottom plate has cooling water lines soldered to it in order to transfer the heat pumped from the top surface to the 10°C cooling water. The thermoelectric cooler was connected to a power supply and the wafer temperature could be controlled from $-23°C$ to 100°C by varying the magnitude and direction of current flowing through the thermoelectric module. The wafer temperature was measured using a k-type thermocouple positioned in a groove between the cover plate and the wafer.

This example deposition system could be reconfigured to run with just a mechanical pump and cold trap. This however would only allow for base pressures in the 10^{-3} Torr range. For example, the mechanical pump could be hooked up in line with the Trap 1 and Trap 2, the diffusion pump, and the QMS could be taken out of the system.

9.   Commercially Available Deposition Systems and Coating Services

There are a number of corporations that currently not only provide deposition (or coating) services but also manufacture parylene deposition systems for sale. Some of the companies that provide coating services include Advanced Coating (advancedcoating.com), Paratronix (paratronix.com), Para Tech (www.parylene.com), Specialty Coating Systems (scscookson.com), Parylene Coating Service (parylencinc.com),

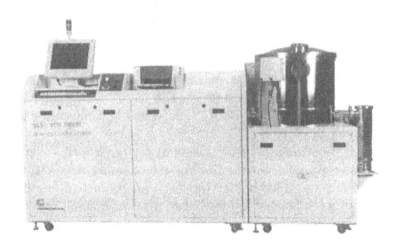

Figure 2.7: **Specialty Coating Systems PDS 2060PC production parylene deposition system, used with the per-
mission of Specialty Coating Systems, Indianapo-
lis, Indiana.**

and Parylene Engineering (paryleneengineering.com). A number of these
companies, including Para Tech, Paratronix, and Specialty Coating Systems manufacture and sell deposition systems. The systems run from
bench top research grade systems through production volume with large
deposition chambers. For example a production level system from Specialty Coating Systems is shown in figure 2.7. According to the SC-
SCookson.com website, this system features a modular generation unit
construction with interchangeable chamber modules and closed loop
monomer pressure control. It also comes with continuous process monitoring, a low-noise direct drive vacuum pump, a message display, a
battery backup and data logging, and easy-access foldout electrical and
instrument panels. Option include an automated vaporizer valve, an
uninterruptible power supply, a load door and chamber lid interlocks,
and a tumble coater chamber module.

Chapter 3

STEP-BY-STEP GUIDE TO DEPOSITING PARYLENE

In this chapter we will present a step-by-step procedure that one would use to deposit parylene using a typical deposition system. The steps for your deposition will most likely deviate from these but they are not meant to be followed exactly but more to give the general understanding of the process of parylene deposition.

1.   Equipment Preparation

All equipment should be in working order and properly cleaned. All wiring and electrical connections should be within code and plugged in the proper receptacle and grounded correctly. If using a diffusion pump, there should be a safety switch such that if the cooling water stops the diffusion pump turns off. The sublimation and pyrolysis furnaces should be clean and calibrated. All heater tapes should be wrapped on the lines per the manufacturers specifications. The cold trap should have been removed and cleaned from the previous deposition. The pumping system should have an adequate amount of oil or fluid and, if using a belt drive mechanical pump, the belt tension should be set correctly. All pressure sensors should be in working order and calibrated. If the deposition chamber is equipped with some form of tumbler this should be inspected for function. All equipment should be maintained in a preventative manner.

2.   Substrate and Source Preparation

Substrate preparation can range from degreasing to a RCA clean depending on the substrate. In general, the surface of the substrate needs to be as clean as possible to promote good bonding to the parylene.

Also, substrates should not have any material with a high vapor pressure that will come off of the substrate and contaminate the chamber. If the substrate needs to be coated all over then it should be situated properly in the chamber or some form of tumbler. For example, when coating ferrites, placing a number of them in a wire "basket" within the chamber that is rotated during deposition causes the parts to tumble and to be coated all over.

The dimer or starting material should be stored as recommended by the manufacturer or supply. Clean dimer should be place in the dimer boat using some form of clean scoop. Use some form of latex gloves when handling the substrates, dimer, and dimer boat to decrease contamination with skin oils. If cleanliness is of great importance then the deposition system should be placed in a clean room.

It is well known that parylene can often not adhere well to surfaces (see section on Adhesion in Chapter 6). If you are using an adhesion promoter carefully follow the manufacturers suggested instruction for applying the promoter to your substrate as well as read the material safety data sheet. Apply the promoter as close as possible to the time of deposition and place the substrate in the chamber for pumpdown as soon as possible after the promoter is applied to insure cleanliness. When using something like A-174 it is possible to set up a delivery system that will allow one to aliquot adhesion promotor vapor into the deposition chamber prior to each run. This vapor will cover the surface of the substrate.

3.   Depositing a Film

Here is a list of steps to follow that will provide a good idea of the process necessary to produce a useful high quality film. Once a good procedure has been established the deposition parameter space should be characterized. Generate a film thickness versus deposition time for a series of pressures. This will provide information on deposition rate as well as identify the lowest and highest pressure that films of good quality can be deposited at. At low pressures the deposition rate will drop to zero, at high pressures the films can get milky in appearance and have very rough surfaces with globules structures.

In preparation for deposition:

- Place the substrate or substrates in deposition chamber.

- Place the appropriate amount of dimer for your deposition in the sublimation furnace and close the entry.

- Initiate pumpdown of the system by either manually turning on the mechanical pump or by hitting the correct button on an automated system.

- Turn on the cold trap system if this is an automated refrigerant or fill the cold trap with liquid nitrogen.

- While the system is pumping down turn on all the instrumentation as any pipe/tube heaters and pyrolysis furnace.

- If you will be heating or cooling the substrate turn this system on now.

- Let the heaters, pyrolysis furnace, and substrate heating/cooling equilibrate at their operating temperature.

- Close the valve between the pyrolysis furnace and the sublimation furnace if the system is equipped with one.

- Turn on the sublimation furnace and let equilibrate at its set point.

To deposit the film:

- Begin opening the valve between the sublimation and pyrolysis tubes.

- Watch the pressure in the deposition chamber rise and begin timing when the pressure gets to the pressure you would like to deposit at.

- Control the position of the valve to maintain the pressure for the duration of the deposition process (in an ideal word the system will automatically perform this function).

- When the appropriate time has passed at the desired pressure, close the deposition valve and shut off all heaters and furnaces.

- Vent chamber, close appropriate valves, and remove substrate after the sublimation furnace has cooled.

After removing the substrate(s) the system can be loaded again for another run. Enjoy the time between runs by reading another chapter in this book or a recently published paper on parylene!

4.   Some Considerations

There are a few items that are worth mentioning: The dimer will stick onto any surface in the sublimation tube that is at a lower temperature than the boat of dimer. Keep this in consideration and use heater tape appropriately to avoid buildup of dimer. You only need to heat slightly higher than the dimer itself and then after the sublimation furnace is turned off you can leave the heaters on longer.

If you are cooling a substrate for a low temperature run it is best to not turn the cooling unit on until after the chamber is at base pressure and the cold trap is filled. If it is cooled before pumpdown when there is moisture in the chamber moisture could condense on the surface and lead to problems with adhesion.

Having a heated valve with variable control between the sublimation and pyrolysis zones can be very useful. Without it, it is much harder to control pressure in the deposition chamber. Make an investment in a heated system that can be in a close loop with the pressure measurement to provide automated control.

Parylene-C has a much higher deposition rate than Parylene-N and Parylene-N has a much higher rate than Parylene-F. In order to get any significant deposition rate with parylene-F the substrate must be cooled to around $0^{o}C$.

It is a good practice to place substrates in multiple positions in the chamber to characterize how deposition rate changes with location. Placing a diffusor in the path of the flowing monomer will also help to reduce thickness nonuniformity.

Chapter 4

PARYLENE-N PRECURSOR CHEMISTRY

1. Dimer

The starting material for the deposition of parylene film is a dimer in the form of a white powder. This dimer can be purchased through a number of companies in the US and abroad, including Para Tech Coating, Inc. (visit www.parylene.com) and Speedline Technologies (visit www.scscookson.com). The dimer for parylene-N, which is also called di-para-xylylene and [2,2]paracyclophane was first reported by Brown as a product of xylene pyrolysis and later was synthesized with high yield by Cram [77]. According to Beach *et al.*, there are two common routes to the synthesis of the dimer, di-para-xylyene; the direct pyrolysis of p-xylene and the 1,6 Hoffman elimination of amine from a quaternary ammonium hydroxide [1]. Beach *et al.* also state that purification of the dimer is often accomplished by recrystallization from the solvent xylene.

The dimer has received considerable research attention due to its interesting strained ring system. The two benzene rings are held parallel to each other and joined via carbon bridges attached at the para positions. X-ray diffraction shows that the conformation of the molecule is one in which the aromatic rings are not planar [78]. Planarity of the benzene rings and the para substituted carbon atoms is not possible due to the repulsion of the π electron clouds between the rings.

Gorham first reported a detailed studied of the quantitative cleaving of the dimer into two monomer units by pyrolysis at temperatures above $550^{o}C$ [13]. This is considered the Gorham process and is the typical process used for producing monomer for parylene deposition. A few researchers have studied converting the dimer to monomer or directly to polymer using radiation. Shimizu *et al.* dissociated dimer into monomer

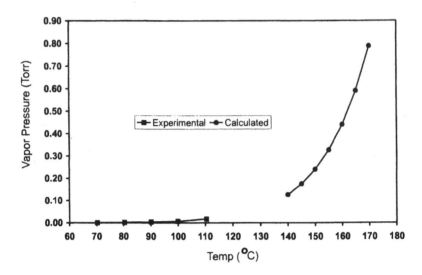

Figure 4.1: The measured and extrapolated dimer vapor pressure from reference [78].

in the gas phase via 193 nm laser photolysis [81]. Mariella *et al.* sublimated the dimer and deposited it as thin layers on a substrate and subsequently converted the dimer directly to polymer by irradiation via deep ultraviolet (< 250 nm) radiation (photopolymerization) [80].

1.1   Thermodynamics

Boyd determined the strain energy of the dimer to be 130 kJ/mole by comparing the experimental heat of formation of the gaseous dimer with that expected for an unstrained reference structure [78]. Nishiyama *et al.* determined the strain energy to be 124 kJ/mole by comparing experimental enthalpies of formation in the gaseous state with the calculated values [79]. The heat of formation of the dimer in the gas phase measured by Boyd was 250.8 ± 8 kJ/mole and that measured by Nishayama *et al.* was 244.1 ± 4.0 kJ/mole.

The vapor pressure of parylene-N dimer has been measured in in the range of T = 70°C to 110°C [78] and the results are shown in figure 4.1. Included in this figure is an extrapolation of the data into the temperature range typically used for sublimation of the dimer during parylene deposition (140°C-170°C). This extrapolation was made using the equation $logP = A/T + B$ and the values for A and B determined by Boyd [78]. In this equation P is the vapor pressure, T is the temperature and A and B are constants that were determined by Boyd experimentally.

1.2   Dimer purity

The dimer purity is stated in detail by some manufacturers and in less detail by others. It is important to know the quality of the dimer that you are using as your precursor. There can be volatile contamination in the dimer that will therefore be present in the deposition chamber during deposition. This could result in incorporation of the contamination into the film as well as a deviation in the expected deposition rate. Of course, depending on the application of the film this may or may not be a problem.

Research has been conducted to examine the magnitude of volatile contaminants present in dimer manufactured by two different dimer manufacturers [45]. This research showed that a large variation in volatile contamination in the dimer can exist between manufacturers. The identity of the contamination was determined by comparing the mass fragmentation patterns of the contaminants to the patterns of possible contaminants.

The magnitude of the volatile contamination was determined in the study [45] by comparing the pressure rise and decay in the deposition chamber for pyrolysis temperatures of $215 \pm 5^{o}C$ and $665 \pm 5^{o}C$. At a pyrolysis temperature of $215^{o}C$, 100% of the dimer flows into the chamber without cracking into monomer [13, 45]. It was seen that the dimer quickly sticks to the $23^{o}C$ deposition chamber walls and does not contribute to the measured pressure of the chamber. Therefore, the chamber pressure at a pyrolysis temperature of $215^{o}C$ is due only to volatile contaminant gases plus the base pressure of the system, which is two to three orders of magnitude lower than the peak pressure due to contamination. At a pyrolysis temperature of $665^{o}C$, 100% of the dimer is cracked into monomer [13, 45]. Thus the chamber pressure at a pyrolysis temperature of $665^{o}C$ is due to the partial pressure of the monomer vapor plus the partial pressure of the contamination gases plus the base pressure.

A contaminant is considered to be any species, excluding dimer, that comes from the sublimation tube. As the sublimation tube is heated up to its set point gases adsorbed on the solid dimer as well as adsorbed on the sublimation tube walls can be desorbed. Along with these adsorbed gases, any liquids or solids volatile at the sublimation temperature that are present will produce vapors that travel out of the sublimation tube into the pyrolysis zone. This includes the dimer, which is a solid at room temperature, and has a significant vapor pressure at $155^{o}C$, see figure 4.1. These gases and vapors travel through the pyrolysis zone and into the deposition chamber. Figure 4.2 compares the pressure in the deposition chamber for two runs with different pyrolysis temperatures. All other

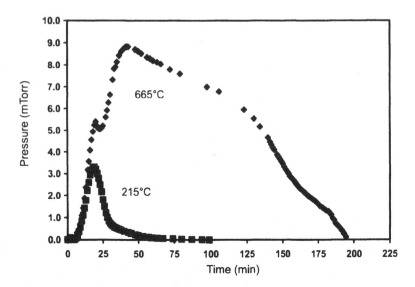

Figure 4.2: Deposition chamber pressure versus time for pyrolysis temperatures of 665°C and 215°C.

parameters are identical for the data in this figure. The sublimation tube is turned on and set to 155°C at t=3 minutes. At 215°C, 100% of the dimer flows into the chamber without cracking into monomer and quickly sticks to the 23°C chamber walls. Therefore, the pressure rise measured in the chamber at this pyrolysis temperature is due only to contamination.

At 665°C, 100% of the dimer is cracked into monomer so the pressure in the chamber is due to the partial pressure of the monomer plus the partial pressure of the contaminants (neglecting base pressure). The magnitude of the contamination can be determined by comparing the two series of data shown in figure 4.2. The data series for 665°C initially shows an increase in pressure as the sublimation furnace heats up. The pressure then reaches a maximum at around 40 minutes and slowly begins to decline. The decline is initially due to the depletion of dimer source and also due to a decrease in the partial pressure of the contamination. After roughly 100 minutes the continued decrease in pressure is due entirely to depletion of the dimer source. The data series for 215°C, showing the chamber pressure due to contamination, initially shows an increase in pressure to a maximum at around 20 minutes. The pressure then quickly decreases and eventually reaches base pressure. Figure 4.3 shows the percentage of chamber pressure for a pyrolysis temperature of 665°C that is due to contamination as a function of time. This figure also shows the data for dimer obtained from a second manufacturer.

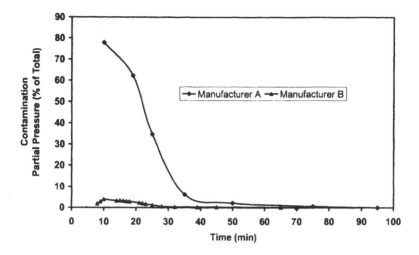

Figure 4.3: **The percentage of chamber pressure due to dimer contamination vs. time for two dimer manufacturers.**

It can be seen that soon after the sublimation tube begins to heat up the pressure in the chamber is mainly due to the contamination for the case of manufacturer A while in the case of manufacturer B there was a much lower amount of contamination present. It took 35 minutes for the partial pressure of the contaminants in manufacturer A's dimer to drop to less than 5% of the total pressure.

Contamination Identification

The identity of the dimer contamination was also examined in the same study via the use of a differentially pumped quadrupole mass spectrometer [45].

Figure 4.4a shows a mass spectrum that was gathered when the pressure in the deposition chamber due to the contamination had reached a maximum.

The mass spectrum of figure 4.4a was compared to the spectra of o-,m-, and p-xylene because of their possible role as contaminants. Figure 4.4b shows the fragmentation pattern of p-xylene gathered with an electron energy of 75 eV and a source temperature of 210°C from the NIST Chemistry WebBook [44]. The inset of the figure shows the structure of p-xylene. Mass spectra of o- and m-xylene from the NIST Chemistry WebBook are nearly identical to that of p-xylene. All of the major peaks seen in p-xylene's mass spectrum appear in the contamination's mass spectrum. Therefore the conclusion was made that xylene was the major contaminant. It was also concluded that there was evidence

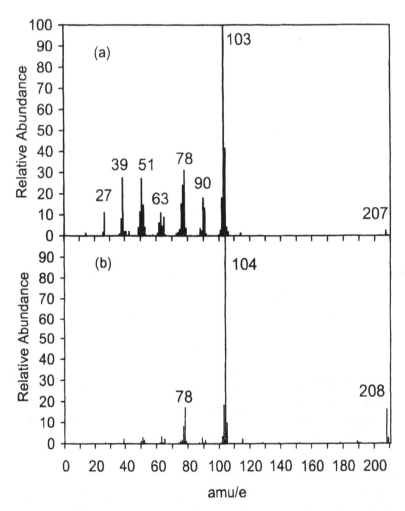

Figure 4.4: Mass spectra for (a) dimer contamination from reference [45] and (b) p-xylene from the NIST Web-Book [44]. Inset in (b) shows the structure of p-xylene.

in the fragmentation pattern that the solvent acetone was present as a contaminant and that it is possible that acetone is used in the dimer purification process.

The Effects of Contamination on Film Processing and Properties

The presence of volatile contamination in the dimer and its presence as a gas in the deposition chamber has the potential to alter the deposition rate and cause it to change with time. The contamination could also become incorporated into the growing film and subsequently cause degradation or changes in the film's properties. Studies have been performed in order to examine these possibilities [150].

The dielectric constant, dissipation factor, leakage current, and thermal stability of the films deposited in this study (one set of films was deposited with a high percentage of contaminants in the chamber and one set with very low percentages) were examined on films of roughly 2200 Å thickness. No difference was found between the two sets. The volatile contamination that was present in the deposition chamber during film growth was therefore either not incorporated into the growing film or was incorporate in such low quantities that the properties of the film were unaffected.

1.3   Mass Spectrometer Fragmentation Pattern

The fragmentation pattern of the dimer was also obtained in reference [45] at a pyrolysis temperature of $215 \pm 5^{\circ}C$. The DPQMS chamber was heated to $200 \pm 5^{\circ}C$ to prevent the deposition of dimer on the walls. The electron energy used was 70 eV.

Figure 4.5a shows the fragmentation pattern of the dimer at a pyrolysis temperature of $215^{\circ}C$. At this temperature no dimer is converted into monomer. The pattern was obtained after the level of contamination had become negligible (roughly less than a few percent). Table 4.1 shows the most probable composition of the main peaks and their parent molecules. Figure 4.5b shows the fragmentation pattern for the dimer from the NIST Chemistry WebBook [44]. The NIST data was gathered at an electron energy of 70 eV and a source temperature of $200^{\circ}C$.

The two patterns displayed in figure 4.5 have many similar peaks, with the experimental pattern containing more peaks than that of NIST. The additional peaks in the pattern can be attributed to increased fragmentation of the dimer. The most noticeable difference between the two patterns is that the NIST spectrum has its largest peak at 104 amu with a smaller peak at 103 amu and the experimental pattern has its largest peak at 103 amu with smaller peaks at 104 amu. The NIST spectrum also has a strong peak at 208 amu with a smaller one at 209 amu and the experimental spectrum has a peak at 207 amu with a smaller one at 208 amu.

2.   Monomer

The monomer, para-xylylene, has been a challenge to study due to its high reactivity [82, 83]. Calculations of the singlet and triplet states

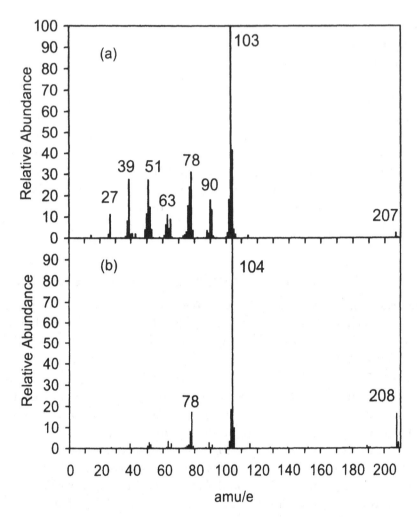

Figure 4.5: Mass spectra for (a) dimer from reference [45] (b) dimer from the NIST WebBook [44].

of the molecule indicate that the molecule is diamagnetic in the lower energy singlet state and paramagnetic in the higher energy triplet state [82]. The singlet state is described by a quinonoid structure and the triplet state is described by a benzoid structure. Benzoid compounds behave as diradicals and exhibit paramagnetism, whereas the quinonoid structure implies the compound is diamagnetic, see figure 4.6.

2.1   Thermodynamics

The energy difference between the singlet and triplet states has been calculated to be 50 kJ/mole [82, 84]. Since the triplet or bi-radical state

Table 4.1: Major peaks appearing in the fragmentation pattern of the dimer shown in figure 4.5 and their corresponding composition.

Amu/e	Composition
27	$C_2H_3^+$
39	$C_3H_3^+, C_6H_6^{++}$
50	$C_4H_2^+$
51	$C_4H_3^+$
52	$C_4H_4^+, C_8H_8^{2+}$
63	$C_5H_3^+$
76	$C_6H_4^+$
77	$C_6H_5^+$
78	$C_6H_6^+$
90	$C_7H_6^+$
91	$C_7H_7^+$
102	$C_8H_6^+$
103	$C_8H_7^+$
104	$C_8H_8^+$

is substantially more energetic it would not be present in any appreciable amount at equilibrium, even at pyrolysis temperatures. Molecular orbital calculations have shown that an equilibrium between the singlet and triplet states significantly favors the singlet state and only a small percentage of the monomer exists as a bi-radical [85]. Calculations of free valence have revealed that the carbon atoms in the para substituted methylene groups have high electron availability [82]. This implies that even in the singlet state the monomer should be extremely reactive, which is why room temperature deposition of parylene is possible.

The heat of formation of the gaseous monomer has been measured using cyclotron resonance and was found to be 209 $\pm$ 17 kJ/mole. It has also been calculated to be 234.8 kJ/mole using molecular orbital techniques [1].

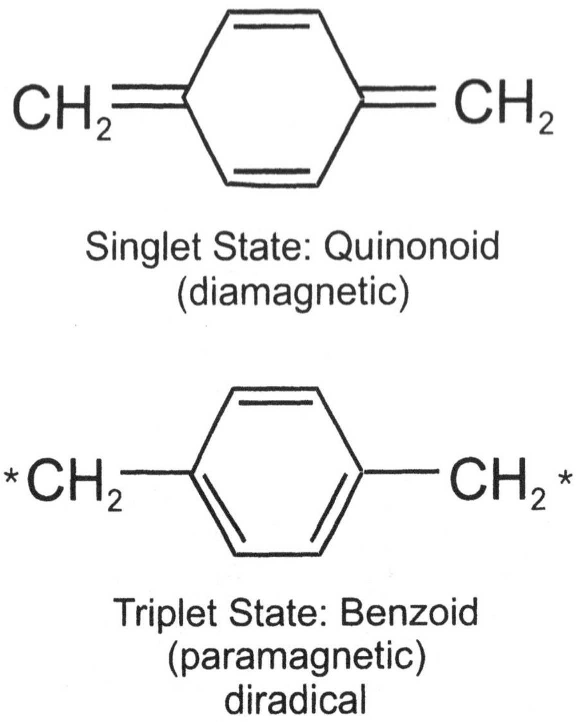

Singlet State: Quinonoid
(diamagnetic)

Triplet State: Benzoid
(paramagnetic)
diradical

Figure 4.6: The singlet and triplet states of the monomer with the asterisks signifying radicals.

The vapor pressure of the monomer has been estimated by Ganguli using the Ambrose group contribution method [7]. This estimation matches up well with the vapor pressure of the solvent p-xylene, which is very similar in structure to the monomer, see figure 4.7.

2.2   Mass Spectrometer Fragmentation Patterns

The fragmentation patterns for the monomer as a function of ionizing electron energy was obtained in reference [45] at pyrolysis temperatures of 665 ± 5°C. These scans are shown in figure 4.8. The mass numbers of the main peaks are labelled in the figure and their most probable composition and parent molecules are given in table 4.2. At 25 eV the only two significant peaks are the monomer $(C_8H_8^+)$ at 104 amu and $C_6H_6^+$ at 78 amu. As the ionization energy is increased the fragmentation as well as the ionization probability increase. An example of this is shown in figure 2.4.

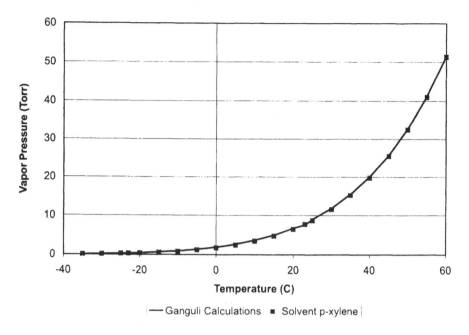

Figure 4.7: **The vapor pressure of the monomer from reference [7] and for xylene.**

3. Dimer-to-Monomer Conversion

The quantitative conversion of dimer to monomer has been studied as a function of pyrolysis furnace temperature in [45]. In this study the temperature ranged from $345 \pm 5°C$ to $665 \pm 5°C$ by starting at $345°C$ and ramping the temperature in $20.0°C$ steps every 12.0 minutes and continuously monitoring the 104 amu monomer peak using a differentially pumped quadrupole mass spectrometer (DPQMS). The 104 amu peak is due only to the presence of the monomer in the deposition chamber. Experiments were performed for quartz, copper (99+%), and nickel (99.9 + %) as liners of the stainless steel pyrolysis tube.

Figure 4.9 shows the relationship between percent conversion and pyrolysis temperature for two experiments performed with quartz, copper, and nickel as liners for the stainless steel pyrolysis zone. It also shows data from Gorham that was obtained by weighing the amount of polymer removed from the deposition zone and dividing that weight by the weight of dimer that was sublimated [13]. It can be seen that for this deposition system (which is the example deposition system), cracking of the dimer begins at $385 \pm 10°C$ and by $565 \pm 10°C$ 100% of the dimer flowing through the pyrolysis zone is cracked into monomer. No significant difference was seen between the quartz, copper, and nickel liners.

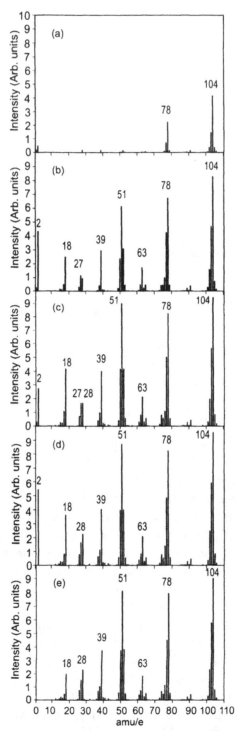

Figure 4.8: Fragmentation patterns for the monomer using
ionizing electron energies of (a) 25 eV, (b) 45 eV,
(c) 65 eV, (d) 85 eV, and (e) 105 eV.

Table 4.2: Major peaks appearing in the fragmentation pattern of the monomer shown in figure 4.8 and their corresponding composition and parent ion.

Amu/e	Composition	Parent Ion
2	H_2^+	Monomer?
18	H_2O^+	Water
27	$C_2H_3^+$	Xylene
28	$C_2H_4^+, N_2$	Monomer, Nitrogen
39	$C_3H_3^+, C_6H_6^{++}$	Monomer
50	$C_4H_2^+$	Monomer
51	$C_4H_3^+$	Monomer
52	$C_4H_4^+, C_8H_8^{2+}$	Monomer
63	$C_5H_3^+$	Monomer
77	$C_6H_5^+$	Monomer
78	$C_6H_6^+$	Monomer
102	$C_8H_6^+$	Monomer
103	$C_8H_7^+$	Monomer
104	$C_8H_8^+$	Monomer

Error bars are shown only for the quartz liner to avoid cluttering the data. The error in percent conversion for the other liners is the same as that for the quartz.

It is important to note that the residence time of the dimer in the pyrolysis zone affects the conversion to some degree, therefore, this curve should only be used as a reference and not considered directly applicable to every parylene deposition system. The pumping speed of your deposition system and flow rate of the monomer can be calculated, as was shown earlier for the example deposition system.

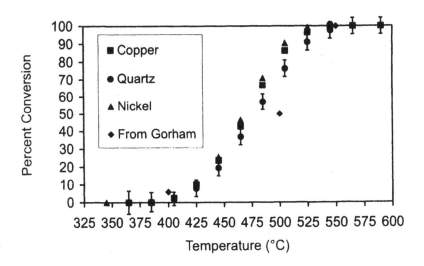

Figure 4.9: Percent conversion of dimer to monomer vs. pyrolysis temperature for multiple tube linings.

Chapter 5

DEPOSITION KINETICS FOR POLYMERIZATION VIA THE GORHAM ROUTE

1.   Introduction

It is known to those familiar with parylene that it is has very unique deposition kinetics. The deposition rate increases as the substrate temperature is decreased, which is the opposite of most CVD reactions. The deposition rate increases as the monomer pressure in the chamber is increased, which is more typical of a CVD reaction. It is also known that there is a minimum pressure below which no deposition occurs. This pressure increases (decreases) as substrate temperature increases (decreases). Before more information is presented on parylene kinetic models it is important to have an overview of CVD kinetics in general.

2.   Overview of CVD Kinetics

A study of the chemical kinetics of a CVD process is particularly important for both an understanding of the process and an optimization of its controlling parameters, with the goal being to obtain a film with the desired properties at the desired deposition rate. Many texts have reviewed the kinetics of CVD processes [89, 90, 91, 92]. This overview presents the general characteristics of a typical CVD process and describes the three types of control which are mass transport, diffusion, and kinetic.

In a typical CVD process the deposition of a thin solid film occurs via a reaction on the substrate surface,

$$A(g) + B(g) \rightarrow C(s) + D(g), \qquad (5.1)$$

where A and B are the gaseous reactants, D is a gaseous product, and C is the solid deposited material. In a CVD process film deposition proceeds through a number of steps: (1) transport of the reactants into the deposition chamber; (2) diffusion of the reactants from the region above the substrate through any boundary layer that may exist; (3) adsorption of the reactants onto the substrate; (4) surface chemical reaction; (5) surface migration and lattice incorporation; (6) reaction products desorption; (7) diffusion of the products away from the surface through any boundary layer that may exist; (8) transport of the products out of the deposition chamber. Notice that step (5) could occur prior to (4) as well as after (6).

These 8 steps are the common steps of a CVD reaction and for any given reaction they occur in series. This means that the reaction rate is controlled by the slow step, the rate limiting step, in the series. The control type depends on the rate limiting step and is one of the three types: mass transport, diffusion, or kinetic.

A CVD process is considered to be under mass transport control if steps (1) or (8) are the rate limiting or slow steps. In this case the reaction could proceed at a faster rate if either the reactant had a higher flow into the chamber or the products had a higher flow out of the chamber. In a mass transport controlled CVD reaction the rate is lower than what could potentially be obtained if the deposition were proceeding under the other control types for the same system. The process is also 100% efficient because all reactant is converted into film.

A CVD process is considered to be under diffusion control if step (2) or (7) is the rate limiting or slow step. In this case, for example, plenty of reactant is flowing into the chamber but the diffusion of reactant through the boundary layer onto the substrate surface is slow. The reaction efficiency in this process is less than 100% but deposition rates would be higher than for the same system under mass transport control.

A CVD process is considered to be under kinetic control if one of steps (3), (4), (5), or (6) is the rate limiting or slow step. In this situation it is the kinetics of the adsorption/desorption of the reactants or products or the kinetics of the chemical reaction itself that limits the growth rate of the film. The reaction efficiency for this type of control is typically much less than 100%, but the growth rate can also be the highest attainable.

3.   Identifying the Control Type

The control type can usually be determined experimentally by examining the influence of the deposition parameters on the deposition rate of the film. The deposition parameters that can be studied include the gas flow rates, substrate temperature, reactant partial pressures, crystallographic substrate orientation, geometrical substrate orientation, and substrate surface area. In order to accurately check the effects of one parameter all others must remain constant.

A change in any one of these parameters (while holding the others constant) could increase, decrease, or have no effect on the deposition rate. For example, increasing the flow rate of the reactants could increase the deposition rate. This could be interpreted as either a mass transport or a diffusion controlled process. Unfortunately, often more than one variable needs to be examined before a conclusion can be made as to the type of process control. A detailed analysis of how a parameter should effect the deposition rate for each of the control types is given in reference [90].

4. Identifying the Control Type for Parylene CVD

The parylene process is slightly different from that shown in equation 5.1 in that there is only one gaseous reactant and no gaseous products. The initiation reaction is believed to take place when a minimum of three monomer molecules join to form a diradical oligomer [46] according to the equation

$$3M(g) \rightarrow P^*_{m=3}(s), \tag{5.2}$$

where M is the monomer, P is the polymer (or oligomer in this case), m is the number of monomer units in the chain, and the $*$ refers to a di-radical. After initiation has taken place, the polymer chain can grow by propagation reaction according to the following equation:

$$M(g) + P^*_m(s) \rightarrow P^*_{m+1}(s). \tag{5.3}$$

The amount of monomer consumed in propagation reactions is 2 to 3 orders of magnitude greater than that consumed in initiation reactions. This leads to the high molecular weight of parylene on the order of 200,000 - 400,000 $gmol^{-1}$, or about 2000-4000 units per chain length [46].

The steps leading to parylene deposition are, based on the reaction mechanism: 1) transport of the monomer into the deposition chamber; (2) diffusion of the monomer from the region above the substrate through any boundary layer that may exist; (3) adsorption of the monomer onto the substrate; (4) surface migration and possibly bulk diffusion of monomer; and (5) chemical reaction (initiation or propagation). Also note that desorption of monomer anytime after adsorption is possible. Figure 5.1 displays the near surface of a film at one instant in time during a deposition process.

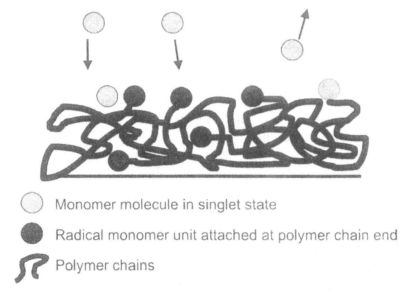

Figure 5.1: **The deposition process for parylene family polymers, not to scale.**

The parylene CVD process has been experimentally determined to be
kinetically controlled [46, 1, 47, 48, 150]. This conclusion was made by
studying the deposition process for deposition parameters of substrate
temperature between $-176^{o}C$ to $+100^{o}C$ and monomer pressures of up
to a few hundred millitorr, depending on the study. Temperature and
pressure are the two variables controlled in the parylene deposition process and their effect on the deposition rate and how this information is
used to determine the control type is discussed below.

4.1 The Effect of Temperature on Deposition Rate

The effect of changing substrate temperature on the deposition rate
as well as on film properties has been studied by many researchers [93,
94, 7, 37, 95, 48, 150]. All reports show an increase in deposition rate
with a decrease in temperature. For the example deposition system this
relationship is shown in figure 5.2.

This experimental data points to a particular control type. For mass
transport control the rate would only be controlled by the transport of
monomer into the system and therefore would not be expected to be a
function of temperature. Diffusion of reactants through a boundary layer
is a temperature activated process and varies roughly as T^2. This would

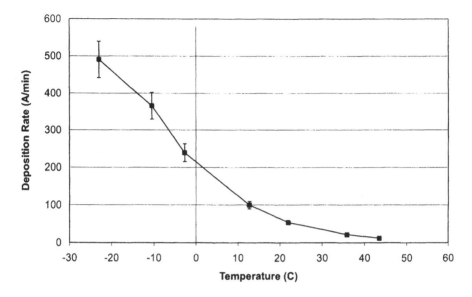

Figure 5.2: Deposition rate vs. temperature for the example deposition system at a pressure of 4.0 mTorr.

mean that for a diffusion controlled process the deposition rate would increase with an increase in temperature. So this leaves the kinetically controlled process as the only alternative.

4.2 The Effect of Pressure on Deposition Rate

The effects of pressure on deposition rate at multiple temperatures has also been studied by many researchers [93, 94, 7, 37, 95, 48, 97, 150, 98]. These studies showed rates that were proportional to P^1, $P^{3/2}$, and P^2. This general power law behavior can also be explained in terms of a kinetic (adsorption) limited process. Deposition rate vs. pressure for the example deposition system is shown in figure 5.3 for substrate temperatures of 22^oC and -23^oC. This data is best fit as proportional to P^1.

4.3 Summary

The kinetic data gathered to date shows that the deposition rate increases as the substrate temperature is decreased and as the monomer pressure is increased. It also shows that there is a minimum pressure below which no deposition occurs. This pressure increases as substrate temperature increases. All of these facts leads on to the conclusion that the deposition of parylene is kinetically controlled by steps 3, 4, or 5 of the parylene deposition process (3=adsorption of the monomer onto the

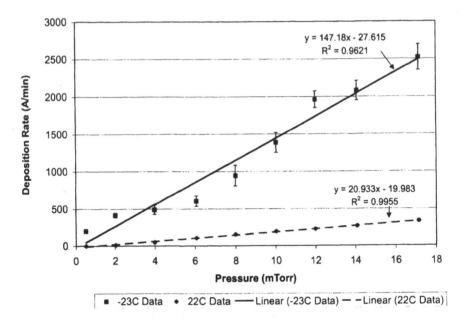

Figure 5.3: Deposition rate vs. pressure for the example deposition system at temperatures of $-23°C$ and $22°C$.

substrate, 4=surface migration and possibly bulk diffusion of monomer, 5=chemical reaction (initiation or propagation)).

5. Kinetic Modeling of Parylene Thin Film Growth

5.1   Introduction

There are a number of models that have been developed to describe parylene deposition and to predict the deposition rate as a function of monomer pressure and substrate temperature [46, 47, 48, 98]. A majority of these models are based on the kinetics of the reaction process, step 5 of the CVD steps, (including both initiation and propagation reactions). The models can be broken down into four types: the first three are those that define the monomer concentration at the surface of the growing film using Flory-type adsorption, Langmuir-type adsorption, or Brunauer-Emmett-Teller (BET-type) adsorption, and the fourth type, having a unique approach, describes and models the deposition process like a chemisorption event. The first three model types have also been presented to include either surface reaction only or surface and bulk reaction (which assumes some degree of monomer diffusion into the bulk of the film during growth). The forth model has been termed the Chemisorption model.

The models previous to the Chemisorption model treat the deposition process as one under kinetic control and are expressed as the product of a kinetic term raised to a power (the first term) and an adsorption, or coverage, term raised to a power (the second term) as is shown below for Beach's model.

The first model ever developed was developed by Beach [46] and includes both surface and bulk reaction. Beach's equation for the deposition rate is given as:

$$R_d = \left(\frac{2k_i k_p D_f^2}{3\rho^3} \right)^{1/4} (C_{M,S}^F)^{3/2}, \qquad (5.4)$$

where k_i, k_p, and D_f are the initiation rate constant $(cm^6/g^2 s)$, the propagation rate constant (cm^3/gs), and the diffusion constant (cm^2/s) of monomer through the polymer film, respectively. The concentration at the surface of the growing film is given by Flory [99] according to the following equation:

$$C_{M,S}^F = \frac{\rho P}{K_H P_{sat}}, \qquad (5.5)$$

where $C_{M,S}^F$ is the surface concentration in g/cm^3 of the adsorbed monomer, ρ is the density of the film, P is the pressure in the deposition chamber, K_H is a dimensionless constant with a value near 3, and P_{sat} is the equilibrium vapor pressure of the monomer at the given temperature. This equation links the deposition pressure to the surface concentration. This expression is borrowed from solution chemistry and it is the limiting form of Henry's law when the concentration of diluent (in this case the monomer) is expressed as a volume fraction [46].

Like Beach's model above, all the models prior to the Chemisorption model are complicated in that they contain terms for the kinetics of the initiation reaction, propagation reaction, and, in some cases, the diffusion of monomer into the bulk. The validity of the models can unfortunately not be verified by analyzing the values of the fitting parameters, as the parameters contain lumped or effective kinetic terms. A detailed analysis of all the models presented in literature to date can be found in reference [150]. The Chemisorption model will be presented here due to its unique approach and merit.

5.2   The Chemisorption Model

Explanation of Model

All of the kinetic models for the deposition of parylene, except for
the Chemisorption model, rely on the form of the surface coverage term
in order to fit the data (see Beach's model equation above). It is this
coverage term that dominates the equation and allows the equation to
fit the experimental data over the appropriate range. The deposition
rate as a function of temperature for CVD parylene is not consistent
with a typical kinetically controlled CVD reaction, which would show a
deposition rate that increases with temperature due to overcoming the
positive activation energy for reaction [89, 90, 91, 92]. Note, however,
that Errede saw a typical thermally activated parylene growth in solution
polymerization [20]. The chemical vapor deposition of parylene is clearly
different.

Any model developed to describe the kinetics of a CVD process should
just describe the kinetics of the limiting step. This rate limiting step is
taken to be the absorption process in the Chemisorption model, and it
turns out to be very fruitful. There are also many other known CVD
processes that are adsorption rate limited [96].

The Chemisorption model is based on the fact that the maximum
deposition rate for any CVD process can be given by (from references
[89, 103, 98])

$$R_d = \frac{SPN_aV_m(60x10^{10})}{(2\pi m_r RT_o)^{0.5}}, \tag{5.6}$$

where the quantity $PN_a/(2\pi m_r RT_o)^{0.5}$ is the flux of the reactant to
the substrate surface in *collisions*/$(m^2 sec)$, P is the pressure in Pa, N_a
is avogadro's number, m_r is the molecular mass in $kg/mole$, R is the
Rydeberg gas constant, and T_o is the temperature of the gas in Kelvin.
The quantity V_m, which is the volume of one molecule in m^3, converts
the flux into units of deposition rate in m/sec, and the multiplication
factor of $60x10^{10}$ converts the units to Å/min. The parameter S is the
fraction of molecules that react after striking the surface, also called the
sticking coefficient here [104]. Without the correction factor, V_m, this
equation is just the rate of adsorption [103].

The mathematical form of the sticking coefficient, which is dependent on temperature, is determined by the energetics of the adsorption
process which can be expressed in the form of a chemisorption type
of potential [103, 104]. This potential for molecular chemisorption is
shown in figure 5.4. This surface potential energy exhibits both a deep
chemisorption well and a shallow physisorption well somewhat further
from the surface. From this picture, an incoming molecule can first be

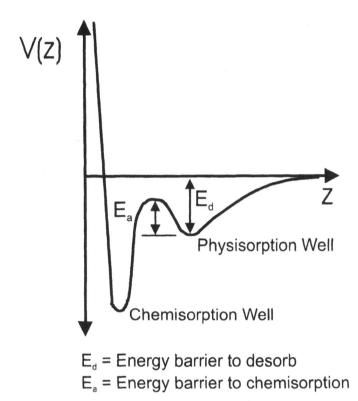

E_d = Energy barrier to desorb
E_a = Energy barrier to chemisorption

Figure 5.4: A molecule-surface potential diagram for a molecular chemisorption.

trapped via physisorption in the outer well, termed a precursor state, and then enter the deeper well at a later time via chemisorption. The sticking coefficient here refers exclusively to the chemisorbed molecules.

In the situation of parylene CVD this can be viewed as the monomer first being physisorbed on the surface with no activation energy (which is typical of physisorption) and then surmounting an energy barrier to chemisorb. The chemisorption would be the reaction of the monomer with a chain end in a propagation reaction and the barrier would be the activation energy for propagation. It is important to note that, in this case, each chemisorption step is equivalent to a propagation reaction between the monomer and a radical chain end and therefore each chemisorption reaction produces a new chemisorption site. This is not typical of a chemisorption reaction.

The term coverage, θ, needs to be clarified in this situation. In normal adsorption situations, coverage refers to the fraction of sites which are not active, or which have had a molecule adsorb. In a typical chemisorption case when a gas is lead to a reactive surface the initial coverage is zero and it will increase with time depending on the temperature, pressure, and energetics until an equilibrium is reached. In the case of parylene CVD, for every molecule that chemisorbs a new chemisorption site, or a new radical chain end, is formed. This means that under steady state deposition, where the deposition rate is constant, the coverage is constant with time. It is easier to think in terms of $(1 - \theta)$, which is the fraction of the surface sites that are reactive, or are radical chain ends. Under steady state conditions this fraction is constant with time. In the model, $(1 - \theta)$ is considered to be constant over the values of deposition parameters analyzed. What this says is that the concentration of reactive chain ends at the surface is fairly constant over the range of deposition conditions considered by the model. This assumption is based on the fact that the chain end density in the film is related to the average chain length (or molecular weight) and this has been determined to remain roughly constant between -176°C and 26°C [102].

In order to develop an expression for the sticking coefficient, S, as a function of coverage a few assumptions are made [104]. First, every site on the surface can support physisorption. Second, adsorbed molecules can roam across the surface in search of unoccupied chemisorption sites to occupy. This means that after a molecule has physisorbed it can chemisorb with a probability P_a, desorb back into the gas phase with a probability P_d, or migrate to an adjacent physisorption well with a probability P_m. The sticking coefficient can be found by adding up the probability of chemisorption at each site a molecule visits [105]. The resultant expression is given as

$$S = S_o \left(1 + \frac{\theta}{1 - \theta} K \right)^{-1}, \tag{5.7}$$

where

$$K = \left(\frac{F_a + F_d + F_m}{F_a + F_d}\right) \times \left(\frac{F_{d'}}{F_{d'} + F_{m'}}\right), \tag{5.8}$$

and F_a, F_d, F_m, are the frequency of chemisorption, desorption, and migration with the prime signifying a molecule that is on a covered site. The frequencies are equal to the respective probabilities multiplied by the average time spent on each site [105].

The limit of the sticking coefficient at zero coverage, or 100% active sites, S_o, is defined as $P_a/(P_a + P_d)$b. These probabilities are given as $P_a = V_a exp(-E_a/RT)$ and $P_d = V_d exp(-E_d/RT)$, where V_a and V_d are pre-exponential constants, E_a is the activation energy for chemisorption, E_d is the activation energy for desorption, and T is the temperature of the substrate, see figure 5.4. So, the expression for S_o reduces to:

$$S_o = \frac{1}{\left(1 + Ve^{-(E_d - E_a)/RT}\right)}. \tag{5.9}$$

The isotherm most commonly applicable to adsorption kinetics is the Langmuir isotherm. This isotherm was shown to apply to parylene deposition in the Chemisorption model. In the case of Langmuir type adsorption the sticking coefficient is given as $S = S_o(1 - \theta)$. Note that this arises because in the Langmuir case the probability for migration, P_m, is taken to be zero, making $K = 1$. Using the Langmuir isotherm, the final form of the rate equation then becomes

$$R_d = \frac{S_o(1 - \theta)PN_aV_m(60x10^{10})}{(2\pi m_r RT_o)^{0.5}}. \tag{5.10}$$

It is interesting to think what this means for parylene deposition. Is seems that if the model works well for $K = 1$, then a majority of the polymerization takes place when a monomer lands right on (or very close to) a reactive site and if an impinging monomer does not land on or near this site it is desorbed. This may be a combination of the fact that the molecules are so large that their mobility is low and also that there is a very low density of sites on the surface (roughly 1 in 1000, see table 5.1) such that even slight surface migration does not significantly increase the probability of finding a reactive site.

Application and Fit to Experimental Data

The final rate equation, equation 5.10, was fit to experimental data by varying three parameters: (1) the value of $(1 - \theta)$, taken to be a constant; (2) ΔE, which is equal to $E_d - E_a$; and (3) V, which is equal to V_a/V_d. The volume occupied by one monomer was calculated from the density of the film and the molecular mass. Also, in calculating the impingement rate, the temperature of the monomer (T_o) was taken to be $298K$, or room temperature, for all deposition conditions. The best fit values for the fitting parameters are shown in table 5.1 and the fit to the experimental data is shown in figure 5.5. The sticking coefficient as a function of temperature was calculated from the results and is shown in figure 5.6.

Conclusion of Model's Validity

The first step in determining the model's validity is to verify that it fits the data well, which it does better than any of the other models as shown in reference [150]. The second step is to check the values of the fit parameters to see if they are reasonable.

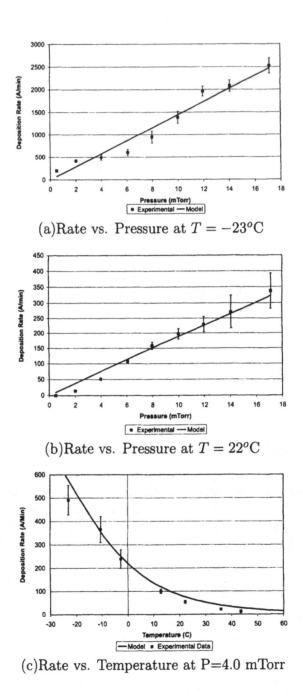

(a)Rate vs. Pressure at $T = -23^oC$

(b)Rate vs. Pressure at $T = 22^oC$

(c)Rate vs. Temperature at P=4.0 mTorr

Figure 5.5: The fit of the chemisorption model to experimental data from the example deposition system.

Table 5.1: Values of the fitting parameters for the chemisorption model

Parameter	Best Fit Value
$(1 - \theta)$	$1.29x10^{-3}$
ΔE	$39.4 kJ/mole (9.41 kcal/mole)$
V	$1.20x10^8$

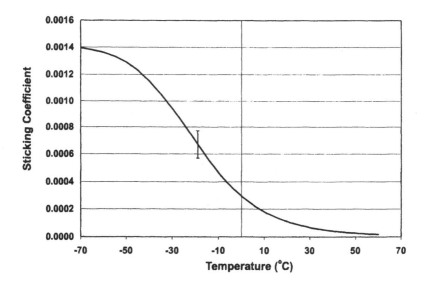

Figure 5.6: The sticking coefficient vs. temperature from the chemisorption model.

The value of $(1 - \theta)$ was found to be $1.29x10^{-3}$, remember this is the fraction of surface sites that are reactive (so roughly one out of one thousand sites is reactive). A value of $6.5x10^{-3}$ to $8.6x10^{-3}$ was calculated based on the estimated value of the average chain molecular weight of 250,000 to 400,000 g/mole [46, 1] and considering the chain ends to be distributed evenly throughout the film. The best-fit value matches up to the calculated value within reason.

The value of $\Delta E = E_d - E_a$ can be also be checked using known data. Considering the activation energy, E_a, to be $8.7 kJ/mole$ as determined by Errede [20], the value of E_d would be around $75 kJ/mole (18 kcal/mole)$. E_d is the heat of adsorption, the heat given off during physisorption.

With this in mind, a typical heat of adsorption for a small molecule like CO or H_2O is typically on the order of $42kJ/mole(10kcal/mole)$ [96, 103] whereas for a larger molecule, like C_6H_6 the heat of adsorption is around $63kJ/mole(15kcal/mole)$ [107]. The heat of adsorption typically increases with molecular weight, for example, it would be expected to increase approximately proportional to molecular weight for a homologous series like the paraffins. The heat of adsorption for p-xylene (C_8H_{10}), a solvent with a very similar structure to the monomer, has been calculated to be $82kJ/mole(20kcal/mole)$ and determined experimentally to be between $70kJ/mole(17kcal/mole)$ and $114kJ/mole$ $(27kcal/mole)$ for adsorption onto a zeolite [108]. Therefore, the value for ΔE also appears to be within reason.

Finally, the sticking coefficient, shown in figure 5.6, also is in a sensible range of values. Parylene is known for its excellent conformallity to substrates. It has been shown to deposit deep into substrate gaps and under overhanging sub-micron structures [5, 7, 1]. This can only occur for a deposition process with a low sticking coefficient.

The results of Errede [20] can now be more fully understood. In solution polymerization there would not be a physisorbed state, there would only be a "chemisorbed" state. Therefore there would only be one activation energy, E_a, and the result would be an increase in growth rate with an increase in temperature.

It is important to note that the chemisorption model fails to predict the correct deposition rate at very high temperatures and very low pressures. This is because, under these conditions, the initiation becomes the limiting step. Parylene family polymers are known to have a ceiling temperature above which deposition will not take place [13]. In reality this ceiling temperature is also dependent on pressure. This is believed to be due to the lack of initiation under these circumstances, probably due to a limited coverage of monomer molecules. This same concept holds true for low pressures, for example, at 25^oC the deposition rate in the example deposition system is zero at 0.5 $mTorr$, while the model predicts it to not go to zero until 0 $mTorr$. The substrate can also play some role in this initiation reaction [106]. A model incorporating the initiation reaction would potentially be a complex one considering all the energetics involved, but it is not necessary to incorporate the initiation reaction to account for most of depositions performed under typical deposition parameters, because the initiation isn't the rate limiting step.

The Chemisorption model does not explicitly include any parameter for diffusion of the monomer into the bulk of the film, as previous models have with limited success. This is not to say that there isn't some degree of monomer diffusion into the bulk, however small. The diffusion of monomer into the bulk was found to be an important concept in the explanation of the surface roughness evolution of deposited parylene films, as will be discussed in a later section [109].

6. Conclusion

Any model developed to describe the kinetics of the deposition process should fit the data over the appropriate range for multiple systems and be a reasonable theoretical interpretation of the process. In the case of parylene CVD, the model should fit data for deposition temperatures from around −40°C up to the temperature at which rates are no longer significant (depends on parylene type) and pressures up to a few hundred mTorr. Outside of these regions the films become porous and thus the density is lowered and film growth rates based on the typical density of the film (about $1.11 g/cm^3$) are no longer valid.

Overall, it appears that the Chemisorption model works well in predicting the deposition rate as a function of pressure and temperature for the steady-state deposition of parylene-N. It also nicely relates the chemisorption rate to the growth rate. The fact that the best-fit values for the fitting parameters are sensible even further justifies this model's validity.

This model should also apply to other parylene family polymers with the values of the fit parameters adjusted based on the energetics of the chemisorption like potential for the monomer-polymer system being described. For instance, with parylene-C, the barrier to chemisorption should be the same as for parylene-N as the polymerization energetics shouldn't differ much for replacing a hydrogen atom on the ring with a chlorine, yet the barrier for desorption should be higher as the mass of the monomer is higher and there will be stronger interaction with the substrate. Effectively this is saying that $E_d - E_a$ is a larger number for parylene-C than parylene-N. The model would therefore predict a higher deposition rate for parylene-C at a given temperature and this is seen in practice. In the case of the fluorinated parylenes like parylene-F, with the fluorine atoms attached to the aliphatic carbon atoms (those participating in the polymer chain bonding) both the energetics of the chemisorption (polymerization) and the physisorption would be different from parylene-N. Parylene-F has a much lower deposition rate than parylene-N at a given temperature, meaning that the $E_d - E_a$ for parylene-F must be smaller than that for parylene-N.

Chapter 6

FILM PROPERTIES

1.   Overview

In this chapter we will present some details on the properties of parylene family polymers. We will also present data from parylene-N films deposited in the example deposition system as well as the method and equipment used to perform the measurements. This should provide enough information to understand the application space for parylene films.

Table 6.1 gives a list of the typical properties for the most common types of the parylene family polymers. Data exists but is limited for types the two types of fluorinated parylenes (ring fluorination and aliphatic fluorination). The reader should search the literature for more data on those polymers not presented here.

2.   Adhesion

Parylene films in general do not exhibit strong adhesion to substrates (typically 2-3x10^6 Pa [162]), however there are a few techniques that can increase the bond strength to acceptable levels. These methods include, but are not limited to, using silane based coupling agents or using a glow discharge or plasma treatment of the substrate surface. These techniques are presented in more detail below.

The strength of adhesion at an interface between two materials is a function of many factors, including the type of bonding present and the actual substrate surface area (factor in surface roughness) in a given footprint. At the lowest end of the bonding strength scale lies van der Waal interactions and at the upper end lies chemical bonding of the covalent type.

Table 6.1: Typical properties of parylene family polymers [1, 150, 94]

Property	Pa-N	Pa-C	Pa-D
Dielectric constant[a] (1 MHz)	2.66	2.95	2.80
Dissipation factor (1 MHz)	0.001	0.013	0.002
Dielectric strength (MV/cm)	300	185-220	215
Volume resistivity (23°C, 50%RH, ω)	1.4×10^{17}	8.8×10^{16}	2×10^{16}
Surface resistivity (23°C, 50%RH, ω)	1×10^{13}	1×10^{14}	5×10^{16}
Melting point (^{O}C)	420	290	380
Glass transition (^{O}C) [168, 167]	13-80	35-80	110
Linear coeff of expansion ($25°C \times 10^{-5}$, K^{-1})	6.9	3.5	—
Heat capacity (25°C, J/(gK))	1.3	1.0	—
Thermal conductivity (25°C, kW/(mK))	12	8.2	—
Density (g/cm^3)	1.110	1.289	1.418
Refractive Index (in plane)	1.661	1.639	1.669
Tensile modulus (GPa)	2.4	3.2	2.8
Tensile strength (MPa)	45	70	75
Yield strength (MPa)	42	55	60
Elongation to break (%)	30	200	10
Static coefficient of friction	0.25	0.29	0.35
Dynamic coefficient of friction	0.25	0.29	0.31
Hardness (GPa, nanoindentation)	0.6	—	—
Water absorption (%)	0.1	0.1	0.1

[a]Parylene-F = 2.24

Testing the strength of adhesion of thin films to substrates is no easy task and many approaches have been used in literature to devise sample preparation methods and processes that can be used. There is an ASTM standard (D 3359-95a) titled "Standard Test Methods for Measuring

Adhesion by Tape Test" that provides a reproducible standard procedure that can be used [158].

Silane coupling agents can provide a hybrid link between on organic and inorganic materials. These coupling agents can form stable covalent bonds with an organic material and multiple inorganic bonds to an inorganic material. Silane coupling agents have a structure of X-R-SiY3 where X is an organofunctional group and Y is a hyrdolyzable group on silicon. The hydrolyzable group can react directly in the right environment with a hydroxide like M-OH to form bond M-O-Si. The X groups are attached to the silicon via C-Si bonds. There are many examples of such coupling agents including those with the following organofunctional groups: azide, mercapto, diamine, epoxy, methacrylate, vinyl, chloropropyl.

One particular coupling agent that is used successfully with parylene is gamma-Methacryloxypropltrimethoxysilane also known as A-174, formula $CH_2CCH_3COOO(CH_2)_3Si(OCH_3)_3$. The application of silane coupling agents to improve adhesion of parylene to substrates was originally patented by Union Carbide in 1971 (US Patent # 3,600,216). In this patent they disclose using various coupling agents to improve adhesion to organic and inorganic substrates of any form. Substrates especially suited for the process are said to be those metal, glass, or organic substrates having hydroxyl, oxide, or epoxy groups on their surface. The siloxane is required to have an ethylenically unsaturated group bonded to the silicon of the siloxane by a carbon to silicon bond and at least one hydrolyzable group attached directly to the silicon of the silane.

IBM also disclosed in their technical bulletin in 1977 a method for applying certain trialkoxysilanes simultaneously with the deposition of the parylene [159]. The advantage here was the coupling agent they were working with could be placed in solid form in with the parylene dimer and, because it had a higher vapor pressure, the silane would be sublimated first and coat the substrate. Then a co-polymer would be formed and finally pure polymer.

The effects of plasma treatments on the adhesion of parylene on various substrates has also been studies [160]. In this report multiple substrates including organic materials like teflon and metals like nickel were treated with oxygen and argon DC plasmas prior to parylene deposition in the same chamber. Improvements in bond strength of one or two orders of magnitude were found in almost every substrate studied except for those of tantalum and tungsten where no increase was seen. It is thought that the plasma pretreatment may be cleaning the surface, as well as imparting roughness and reactive sites, all of which can increase adhesion.

Other articles have reported results of using a glow discharge to treat a surface before polymerization [161, 162]. The glow discharges studied included argon, oxygen, and methane. In the case of methane it was shown that there is some degree of polymerization of methane on the surface of the substrate and radicals could be produced for covalent bonding to parylene. It was shown that the methane glow discharge had the greatest impact on increasing the bonding strength. For example, in the work done in reference [161], the pull strength of the interface increased and debonding occurred in the parylene layer or in an epoxy layer connecting the parylene sample to a pull test apparatus.

3.   Cystallinity

The parylenes, like many polymers, have varying degrees of crystallinity. The percent crystallinity has been shown to be a function of deposition parameters and post deposition processing [102, 163, 164]. In particular, it has been shown that the percent crystallinity can range from about 20 percent for an as-deposited film to about 80 percent (for parylene-N) after a high temperature anneal slightly below the crystalline melting temperature [164].

The crystalline regions within the polymer can be found in sub- micrometer domains that are dispersed randomly in the amorphous phase. As in many polymers, because the chain length is so high, one chain can participate in multiple crystalline domains. The crystalline domains are more resistive to chemical attack than the amorphous regions and therefore, because of this and the linkages between domains, the polymer will stay together and not dissolve when exposed to rigorous solvents, giving parylene its strong chemical resistance [1].

The crystal structures of parylene-N and -C have been studied in depth [165, 166, 1]. Parylene-N exhibits polymorphism with an α form present after deposition which converts irreversibly (controversial) to a β form when heated to 220°C. The α form has a monoclinic unit cell with a=5.92 Å, b=10.64 Å, c (fiber axis)= 6.55 Å, and β =134.7° and the space group is C2/m [166]. The α and β form produce major x-ray diffraction peaks at $2\theta = 16.79°$, 22.52° and $2\theta = 20.00°$, respectively [167]. Parylene-C exhibits one form of crystalline structure with a = 5.92 Å, b = 12.69 Å, and c = 6.66 Å, with beta = 135.2° [1].

Along with a strong affect on the chemical resistance of the polymer, the crystallinity also contributes to the optical bi-refringence [164]. It has been shown that the optical bi-refringence is a linear function of crystallinity for parylene-C and an aliphatic fluorinated version while it is non-linear for parylene-N.

Table 6.2: Thermal degradation temperatures (^{o}C) of the parylene family polymers [126, 118, 150, 117]

Type	Air	Nitrogen	Vacuum
Parylene-N	175-260	350-490	390-490
Parylene-C	125-260	350-490	—
Parylene-F	400-500	510-530	530

4. Thermal Stability

There have been a number of studies on the thermal degradation of parylene family polymers in air, nitrogen, and in vacuum [117, 118, 119, 120, 121, 122, 123, 124, 126, 127, 128]. These studies have used techniques such as differential scanning calorimetry (DSC), thermal gravimetric analysis (TGA), mass spectrometry (MS), gas chromatography (GC), and thickness change as a function of temperature and time to determine the film's thermal properties, most importantly the decomposition temperature. A good introdctory text on polymer degradation is *Polymer Degradation and Stabilization* by N. Grassie and G. Scott [73].

The reported temperature at which thermal degradation begins for different parylene family polymers has varied slightly in literature and is dependant upon the technique used to evaluate the degradation. Table 6.2 gives the range of values reported for degradation in air, nitrogen, and in vacuum.

There have been a number of studies on the kinetics and process of chain scission due to thermal degradation of parylenes. It has been determined that oxidative chain scission is the main mode of degradation for non-fluorinated parylenes and this limits their use at elevated temperatures in air or in an environment containing oxygen. In one report, the oxidation of parylene-C, -N, and -D was studied between 125°C and 200°C [60]. A correlation between oxygen content and mechanical properties indicated that the amount of oxygen incorporated in the films before significant degradation in mechanical properties (a 50% loss in tensile strength) is about 1000 ppm for parylene-D and about 5000 ppm for parylene-C and -N. The long term continuous use temperatures in air were estimated for 100,000 hours of use to be 57°C for parylene-N, 72°C for parylene-C, and 112°C for parylene-D.

It has also been shown that parylene films can be modified with an anti-oxidant to increase the use temperature in air [128]. In this work films of pure parylene-C and those of parylene-C co-deposited with a sterically hindered phenol were deposited on aluminum substrates and annealed in an air circulated oven from 140 to 200°C. The addition of the anti-oxidant successfully prevented the oxidation of the polymer films for more than 250 hours at 140°C, 5 hours at 180°C, and 1 hour at 200°C. This method was patented in 1979, US4163828.

As is standard with polymers, the amorphous region is characterized by a glass transition temperature (T_g), above which groups on the main polymer backbone have some degree of mobility. The crystalline region can be characterized by its melting point (T_m), above which the polymer turns to a liquid. Table 6.1 gives literature values of T_g and T_m.

5.   Chemical Properties and Biocompatibility

Parylene film are very chemically resistive and are known to be very difficult to remove once deposited. This has probably dissuaded a number of potential users of parylene as a moisture barrier as rework can be a problem. At temperatures below their melting point the polymers are not soluble although they will swell as the solvent absorbs into the bulk of the polymer. Typical swelling resulting from immersion in solvents at room temperature will cause anywhere from 0 to 3% increase in volume. More data on swelling from specific solvents can be found in reference [1].

The biocompatibility of materials to body tissue, body fluids, moisture, electrolytes, etc. is imperative for medical implants and surgical components. Parylene has been investigated for use as a biocompatible electrical insulator for sensors, electronic circuits, and other implanted devices [174, 175, 176, 178, 172, 177, 173, 43]. For example, parylene has been examined for use as an insulator on metal micro-electrodes for stimulating and recording neuronal action potentials [172, 177] and is also widely used as a coating for cardiac pacemakers [173]. Another example is the use of parylene as a corrosion preventing coating of neodymium-iron-boron magnets when they are used intra-orally [43].

Parylene's chemical inertness and its ability to deposit as a uniform coating over complex geometries without pinholes gives it a major advantage over other materials. Other properties that lend to this type of application include its high dielectric strength and its low static and dynamic coefficient of friction.

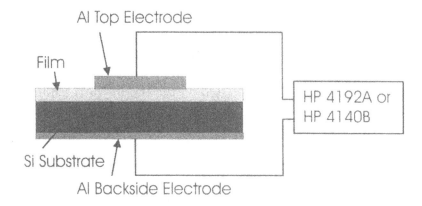

Figure 6.1: The set-up for testing a film's electrical properties.

6. Properties and Measurement Techniques for Parylene-N Films

This section gives details on measurement techniques and results for property measurements of thin parylene-N films. The techniques can be extended to other parylene family polymers as well as other thin films in general.

6.1 Electrical Properties

The electrical measurements on films deposited in the example deposition system were made using metal-insulator-semiconductor (MIS) structures. The films were deposited on highly doped P-type Si substrates (resistivity ≤ 0.020 ohm-cm) and then Al electrodes were deposited on the film's surface via e-beam deposition through a shadow mask with well defined circular apertures. Al was also deposited on the backside to provide ohmic contact to the Si. Direct current leakage measurements were made by connecting to the Al coated backside and the upper electrodes and scanning the voltage on a HP 4140B picoammeter-voltage source from 0 volts to 100 volts at 1 volt increments while recording the current flow, see figure 6.1. The capacitance and dissipation factor were gathered using the same connections to the electrodes and using an HP 4192A LF Impedance Analyzer at a frequency of 1 MHz with an oscillation voltage of 30 mV. All instruments were properly calibrated prior to use and automated data acquisition was performed using MDC CSM/WIN Semiconductor Measurement System software.

Dielectric Constant and Dissipation Factor

The dielectric constant and dissipation factor for parylene-N films deposited in the example deposition system were measured using the procedure and equipment stated above. In order to insure accuracy, four different size electrodes were deposited on the topside of the film. The electrode sizes were defined in the shadow mask using a lithography technique for good edge definition. The actual diameters of the electrodes were measured post deposition using a calibrated measuring microscope. Capacitance and dissipation factor were measured for ten electrodes of each size. Film thickness was measured at the edge of each electrode using a Nanospec tool. The dielectric constant was then calculated from the measurements on each electrode using the following equation:

$$\varepsilon = \left(\frac{Cd}{\varepsilon_o A} \right) \quad\quad (6.1)$$

where ε is the dielectric constant, C is the capacitance in Farads (F), d is the film thickness in meters (m), ε_o is the permittivity of free space and equal to 8.854 F/m, and A is the area of the electrode in square meters. Electrode areas used in this study were 3.14×10^{-6} m^2, 7.06×10^{-6} m^2, 1.96×10^{-5} m^2, and 3.85×10^{-5} m^2. The average dielectric constant was determined to be 2.66 ± 0.02. Note that this dielectric constant is that perpendicular to the plane of the film, or out-of-plane. This value is consistent with typical values reported in literature [1, 122]. The average dissipation factor for the 40 measured electrodes was 0.005 ± 0.001 which is slightly higher than typical values presented in the literature, which is 0.001 [1].

Leakage Current and Breakdown Voltage

Leakage current was measured as described above on 15 electrodes of area = 2.03×10^{-7}m^2 on a 3300Å thick film. Figure 6.2 shows typical results for a sample of electrodes. The average leakage current density was found to be $9.0\pm4.0\times10^{-9}$ A/cm^2 at 2.0 MV/cm. The maximum electric field that could be applied to the film using the testing equipment on hand was 2.98 MV/cm. Only one of the 15 samples showed breakdown under this field so the average breakdown field strength for parylene-N is greater than 2.98 MV/cm. This agrees with other data in literature which states that the average breakdown strength of parylene-N was found to be 3.33 MV/cm [122].

6.2   Mechanical Properties

The hardness and elastic modulus of parylene-N were measured using a nanoindentation technique described in more detail in reference [135].

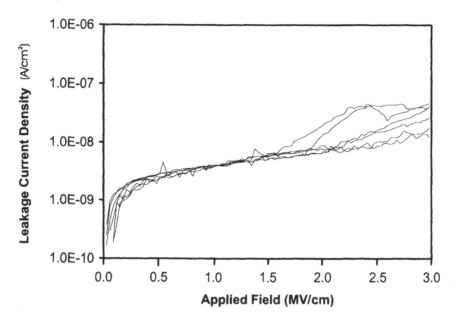

Figure 6.2: Leakage current density versus field strength for parylene-N.

The average hardness for a $10,000\mathring{A}$ thick film was found to be 0.61 $\pm$ 0.03 GPa. This is similar to previous reports of 0.5 GPa and 0.3 GPa [124, 125]. Zhang *et al.* have found the hardness and modulus to be dependent on deposition pressure, temperature, and post anneal treatments [125].

6.3 Optical Properties

The index of refraction, bi-refringence, and film thickness was measured using a J.A.Woolam Co., Inc., $M-44$ spectroscopic (multi-angle, multi-wavelength) ellipsometer. The ellipsometric data are gather for 44 wavelengths from 410 nm to 742 nm for three incident angles, typically 65°, 70°, and 75° from the plane of the film. Once delta and psi values were obtained as a function of wavelength and incident angle a Cauchy model was used to find the Cauchy coefficients for the film (a, b, and c). The refractive index was then calculated for in-plane and out-of-plane cauchy coefficients (where $n = a + b/\lambda^2 + c/\lambda^4$). The film thickness was also a parameter in the model and could be determined with an accuracy of a few $\mathring{A}$ under the best circumstances. Other parameters that were added at times to improve the model were surface roughness and film thickness irregularity.

The index of refraction and bi-refringence of as-deposited parylene-N films (deposited in example deposition system) were measured as described above on multiple film samples of similar thickness. The in-plane and out-of plane refractive index at 634.1 nm was found to be 1.66 ± 0.01 in the plane of the film and 1.60 ± 0.01 perpendicular to the plane of the film. The optical axis was found to be perpendicular to the plane of the film, which is consistent with earlier results [1]. This gives a bi-refringence of 1.60 - 1.66 = -0.06 ± 0.01, which is slightly lower than a previous report of -0.075 ± 0.001 [1]. Senkevich found that the bi-refringence of parylene-N changes as a function of thickness [134].

6.4   Surface Morphology

The surface morphology of a deposited film is a function of the deposition conditions. The final roughness of a film can be an important factor in many applications including those involving adhesion, lubrication, and surface functionality [49, 50]. The knowledge of how the roughness or surface morphology changes with deposition conditions is therefore important.

The surface morphology of thin polymeric films has been examined using atomic force microscopy for films such as polypyrrole, sexithienyl, as well as for plasma polymerized films [51, 52, 53]. The evolution of surface roughness of parylene-N has been studied using an atomic force microscope. The roughness was examined as a function of deposited film thickness for different deposition conditions of temperature, pressure, and time [109].

Thin films grown under non-equilibrium conditions often show scaling behaviors and have attracted considerable attention [110, 111, 112]. The major efforts have focused on studying the growth of deposited metal and semiconductor thin films [110, 111, 112]. So far, only a few works have been performed on polymer thin films and these works show that the such films display various scaling behaviors [51, 53]. Most polymer growth is quite complex and the results are not trivial to interpret when compared to those of metal and semiconductor growth processes.

Films for this experiment were deposited on Si wafers at a temperature of 25°C. Pressure was controlled to within ±0.3 mTorr and deposition time to within 30 seconds. Film thicknesses were measured using ellipsometry and growth rates were extracted from the slope of thickness versus time plots. The deposition pressure was varied from 2 mTorr to 7 mTorr and the deposition rate, R, ranged from 1.95 ± 0.03 to 13.9 ± 0.1 nm/min. Depositions were performed at 7 different pressures within this range leading to 7 different deposition rates.

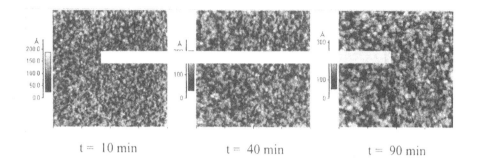

t = 10 min   t = 40 min   t = 90 min

Figure 6.3: The typical surface morphologies (4 x 4 μm^2) measured by AFM for growth time of $t = 10$ min, 40 min, and 90 min at the growth rate $R = 9.5$ nm/min.

The surface morphology was measured using contact mode Atomic Force Microscopy using the Park Scientific Auto CP AFM described. Repeated scans with a small force set point ($= 1$ nN) were carried out in order to ensure no obvious distortion caused by tip-sample interaction. An example of the measured surface morphology for a series of films deposited at a rate of 9.5 nm/min for times of 10, 40, and 90 minutes times is shown in figure 6.3.

Several parameters are required to describe the random rough surface. First of all, the interface width w is defined as the square root of $< [h(\mathbf{r}) - < h >]^2 >$, where $h(\mathbf{r})$ is the surface height at position $\mathbf{r}$, and $< h >$ is the average height. w is the "amplitude" of the surface fluctuations and thus describes the property of the surface at large distance. The lateral correlation length ξ describes the "wavelength" of the surface fluctuations. Two surfaces having the same interface width w and the same lateral correlation length ξ may still possess very different local roughness fluctuations. A common way to describe the local interface roughness is to use the height-height correlation function $H(\mathbf{r})$, which is defined as $< [h(\mathbf{r}) - h(0)]^2 >$. For large r, the surface height fluctuations should not be correlated and $H(\mathbf{r}) = (\rho r)^{2\alpha}$, where ρ is a constant and α is known as the roughness exponent. α describes how "wiggly" the local surface is and has a value between 0 and 1. The height-height correlation function reaches a constant length ξ, beyond which the surface height fluctuations are not correlated.

The height-height correlation function, $H(r,t)$, versus distance, r, is plotted in log-log scale for $R = 5.5$, 9.5, and 13.9 nm/min as shown in

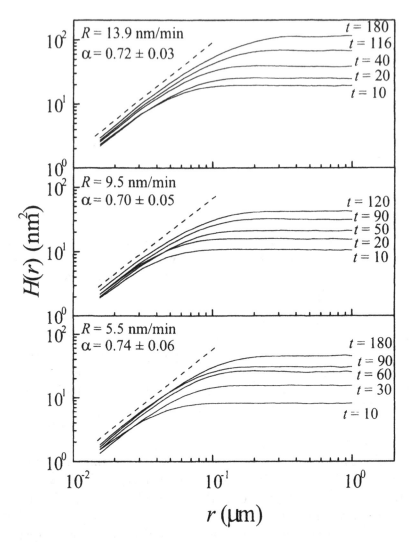

Figure 6.4: **The equal time height-height correlation function** $H(r, t)$ **versus distance** r **for** $R = 5.5, 9.5,$ **and 13.9 nm/min, respectively.**

figure 6.4. The overall behaviors of the height-height correlation function are clearly similar for the three different deposition rates: at small lateral length, the height-height correlation functions do not overlap each other, which implies that the growth is not stationary, i.e., the local slope, ρ is a function of the growth time [114]. The roughness exponent values for the three rates are within experimental errors, $\alpha = 0.74 \pm 0.06$ for $R = 5.5$ nm/min, $\alpha = 0.70 \pm 0.05$ for $R = 9.5$ nm/min, and $\alpha = 0.72 \pm 0.03$ for $R = 13.9$ nm/min. The average roughness from all 7 growth rates

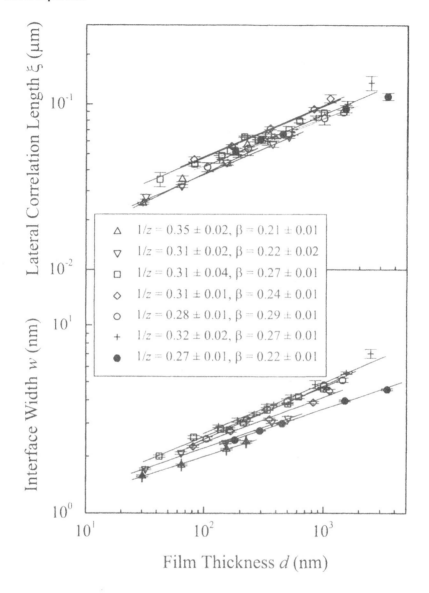

Figure 6.5: The interface width w and the lateral correlation
length ξ versus film thickness d for various growth
rates and substrate temperatures.

studied is 0.72 ± 0.05, however if one considers the tip effect the roughness
exponent could be between 0.5 to 0.7 [115].

The interface width, w, and the lateral correlation length, ξ versus
film thickness, d, are shown in figure 6.5. For different growth rates, the
log-log plots of w versus d are almost parallel, and, for the same thickness

d, the higher the growth rate R, the larger the value of w. The average growth exponent is $\beta = 0.25 \pm 0.03$, which means $w \sim d^{0.25 \pm 0.03}$. In order to determine ξ accurately, the two dimensional autocorrelation function was calculated, $C(\mathbf{r}) = \langle h(\mathbf{r}) h(\mathbf{0}) \rangle$ for each AFM image, and then the quadrant circularly averaged autocorrelation function $C_c(r)$ was used to determine ξ by the relation $C_c(\xi) = C_c(0)/e$. The log-log plots of ξ versus d are almost parallel. The average dynamic exponent obtained is $1/z = 0.31 \pm 0.02$, which means $\xi = d^{0.31 \pm 0.02}$. The data for different growth rates tend to overlap each other which suggests that in this range of deposition parameters the vertical roughness and lateral correlation length of the film growth are independent of the growth rate.

A similar set of experiments was performed at a substrate temperature of -30°C and $P = 7$ mTorr. The growth rate was 53.3 nm/min, and the growth gave a similar scaling behavior: $\alpha = 0.70 \pm 0.02$, $\beta = 0.22 \pm 0.01$, and $1/z = 0.27 \pm 0.01$, see figure 6.5.

The experimental results show that the scaling behaviors of the growth front do not change for a growth rate spanning almost 1 order of magnitude. Of the known theoretical results for dynamic roughening, only the MBE nonlinear surface diffusion dynamics proposed by Lai and Das Sarma gives similar results ($\alpha = 2/3$, $\beta = 1/5$, and $1/z = 3/10$) [116]. However, this MBE model predicts stationary growth with a constant local slope which was not observed in the case of parylene deposition. Chain relaxation during deposition and intermolecular interactions, combined with monomer diffusion into the bulk, could potentially lead to a very different roughening mechanism. Zhao et al. have proposed a roughness evolution model for parylene that includes the possibility of monomer diffusion into the growing bulk film [109]. This model predicts the evolution of roughness and this could be more support to add to the possibility of monomer diffusion into the bulk during film growth.

In summary, the growth front roughness of linear parylene-N films grown by CVD was investigated using atomic force microscopy. The interface width w was found to increase as a power law of film thickness d, $w \sim d^{\beta}$, with $\beta = 0.25 \pm 0.03$, and the lateral correlation length ξ grows as $\xi \sim d^{1/z}$, with $1/z = 0.31 \pm 0.02$.

6.5   Thermal Stability of Parylene-N

The thermal stability of as-deposited parylene-N under vacuum was analyzed on films from the example deposition system using two methods. In one method a thermal desorption spectrometer system was used to monitor the outgassing and gaseous degradation products of the polymer as a function of sample temperature. This method was used to investigate the thermal decomposition products as well as to identify any

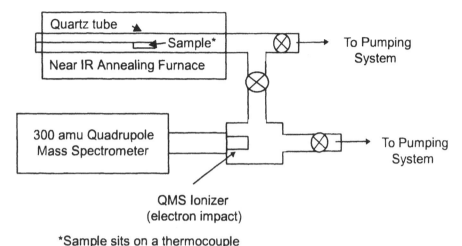

Figure 6.6: A diagram of the thermal desorption system.

volatile materials present in the as-deposited film. The second method used was measuring film thickness before and after 120 minutes of vacuum annealing at a series of temperatures from 300°C to 430°C. This method was used to help determine the temperature at which thermal degradation begins to occur. The results of both techniques will be discussed below.

Thermal Stability via Thermal Desorption Spectrometry (TDS)

A diagram of the thermal desorption system is shown in figure 6.6. The thermal desorption spectrometer (TDS) system consists of a near-IR cold-wall annealing furnace manufactured by Ulvac Sinku-Riko (type MILA-3000) connected to the 300 atomic unit (amu) quadrupole mass spectrometer. The sample under investigation sits in the annealing furnace chamber on a K type thermocouple on a quartz holder. The furnace has a quartz body that allows over 95% of the near-IR light to pass through to selectively heat the sample (parylene on silicon) while not heating the chamber walls.

The sample heating rates and annealing temperatures were automatically controlled via a programmable controller integrated into the annealing unit. The TDS system is pumped using a diffusion pump with a liquid N_2 cold trap. The base pressure during a TDS run and for annealing was in the mid 10^{-7} Torr range.

TDS is a common tool used to study the degradation of polymeric materials [129, 130, 131, 132, 133]. It can provide evidence of reactions (in this case thermal degradation reactions) that are proceeding with the evolution of volatile material. Because of the cold-wall nature of the experimental apparatus, the volatile material that reaches the mass spectrometer is only a fraction of the total volatile material that is given off at the sample temperature. Degradation products of polymers are typically separated into the following fractions: A) products volatile at -196°C, B) products volatile at ambient temperatures but not volatile at -196°C, C) products volatile at degradation temperatures but not volatile at ambient temperatures, and D) solid residue. Fraction A is comprised of simple molecules like hydrogen, carbon monoxide and methane, while fraction B is comprised of molecules with masses up to around 150 amu [73]. The mass spectrometer used on the samples for which data is presented below was measuring degradation products composed of materials from fractions A and B and possibly some C depending on the volume produced and its sticking coefficient on 25°C walls.

The results of the TDS study show two temperature regions were species are seen to desorb or be liberated from the film. As the temperature of the film is increased from room temperature the first peak that is seen is that of 104 atomic mass units (amu) which is the mass of the monomer. This peak begins, depending on the temperature ramp rate and film thickness, in the neighborhood of 150°C, reaches a maximum around 220°C, and decreases to background levels at around 300°C. Then, in the neighborhood of 400°C a peak at 91 amu begins to grow followed by a number of other peaks as the temperature rises. The peaks, which increase as the intensity of outgassing of the film due to degradation increases, reach a maximum intensity at around 510°C. These two separate regions are discussed below.

The first temperature region of interest is that between 150°C and 220°C. In this region the 104 amu peak is seen to rise and fall. This peak was studied in detail on samples prepared under identical conditions and sample size. The results show that the total amount of 104 amu measured, which is proportional to the area under the curve shown in figure 6.7, increases as the film thickness increases. This is evidence that the species responsible for the peak is not just from the surface of the film but also from the bulk, as the surface area is constant. Other researchers have seen volatility of parylene-N of up to 20 weight percent at temperatures below 400°C but did not determine the identity of the volatile materials [119, 123]. These studies concluded that a probable source for the low temperature volatile material might be low molecular weight oligomers.

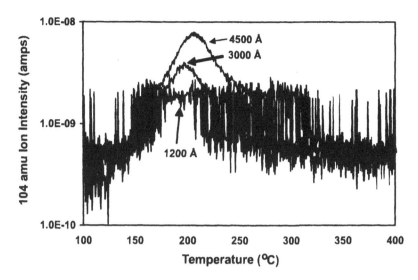

Figure 6.7: The 104 amu ion intensity measured by the TDS system as a function of sample temperature for three film thicknesses at a ramp rate of 7°C/min.

The source of the 104 amu peak could be, 1)monomer trapped in the film, or, 2)larger oligomers (dimers, trimers, etc.) that are volatile in this temperature range that fragment in the RGA to yield the 104 amu main peak. It is believed, as discussed earlier, that there may be some degree of monomer diffusion into the film during deposition [1, 46, 109]. The 104 peak could therefore be evidence of this. No peak larger than the 104 amu peak was detected up to the test limit of 300 amu in this temperature range. This does not rule out the possibility of oligomers fragmenting to produce the 104 amu peak because the parent ion (the oligomer) would probably have a very low ion intensity in the residual gas analysis portion of the TDS system and might not have been above the background level. The maximum peak height of the 104 amu in this temperature range was only about 1 order of magnitude above the noise or background level. Also, as the weight of the oligomer increases the vapor pressure at a given temperature decreases, making it less likely that large oligomers are de-gassing from the film.

In order to investigate the amount of material desorbed in this region that was responsible for the 104 amu peak the following action was taken. Samples were measured for thickness and then taken up to 350°C under vacuum in the TDS unit and the 104 amu peak was monitored. These samples were then removed from the unit and the thickness was re-measured. The initial and final thicknesses were within ±10Å, and,

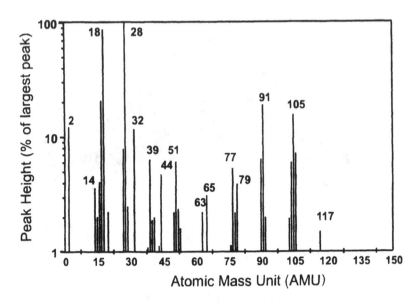

Figure 6.8: The fragmentation pattern for the thermal decomposition products of parylene-N at 510°C.

therefore, no thickness change was detected. This result, combined with the small peak height, lead to the conclusion that the total amount of species producing the 104 amu peak is very small.

The second temperature region of interest is that beginning at around 400°C, the temperature at which decomposition begins. At this temperature a peak at 91 amu begins to rise followed by many others. Figure 6.8 shows the degradation spectrum at 510°C, the temperature at which the peaks reached a maximum. The ion currents are normalized to that of the ion with the largest signal and the major peaks are labelled.

The peaks seen beginning at 400°C are due to the thermal degradation of the film. Chiang used a thermal desorption technique to analyze the thermal stability of parylene-N under vacuum and his results are very similar to those discussed here [124].

The 91 amu peak that was seen in this study was also seen by Chiang and Jellinek [124, 120]. Jellinek attributed this mass to be that of toluene, a decomposition product of the random thermal decomposition of parylene [120]. The 91 amu peak was chosen for more detailed investigation because it is believed to be a parent ion as well as the first mass seen. The 91 amu peak was followed as a function of temperature for samples of the same thickness deposited in the same run. This experiment was carried out at three different ramp rates to compare the effect of ramp rate on the temperature at which the 91 amu peak was first seen

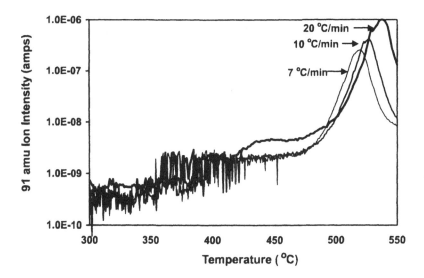

Figure 6.9: The 91 amu ion intensity measured by the TDS system as a function of sample temperature for ramp rates of 7, 10 and 20°C/min, film thicknesses = 4500 Å.

and when it was at its maximum. Figure 6.9 shows the ion intensity of the 91 amu peak measured by the TDS as a function of temperature for heating rates of 7, 10, and 20°C/min. It can be seen that the peak height is clearly above background in all three cases at a temperature of slightly less than 400°C and it reaches the maximum height at successively higher temperatures as the rate is increased. From this data it appears that thermal decomposition begins a temperatures slightly below 400°C. This result was confirmed by other experimentation (see next section).

It is also important to note that no contaminant, or solvent, is seen degassing from the film during TDS experimentation.

Thermal Stability during Isothermal Vacuum Annealing

Although thermal desorption spectrometry is a good tool for comparing degradation products and temperature stability, the data produced is a function of ramp rate as thermal equilibrium is not attained. In order to further verify thermal stability, samples of parylene-N were annealed for 120 minutes at various temperatures from 300°C to 430°C under vacuum. Film thicknesses were measured before and after annealing using ellipsometry.

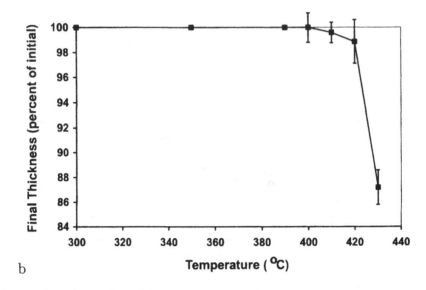

b

Figure 6.10: Final film thickness as a percentage of initial thickness versus annealing temperature.

The results are shown in figure 6.10 where the final thickness (as a percent of the initial thickness) is plotted as a function of temperature. The initial film thickness for these samples was 3300 Å. The samples had a 0.0 ± 1.0 percent thickness loss at $400^{o}C$ and a 0.4 ± 1.0 percent loss at $410^{o}C$. The TDS results show that some decomposition appears to begin at temperature slightly below $400^{o}C$, but this must not be significant enough to see after 120 minutes of annealing. If the annealing time was increased further at $400^{o}C$ it is expected that the film would eventually begin to show a thickness loss.

6.6 Summary

The thermal desorption study revealed that monomer or low molecular weight oligomers exist in the film and are desorbed in the temperature range of $220^{o}C$ to $350^{o}C$. These species exist in very small amounts in the film and their removal from the film did not produce any measurable thickness change. It is expected that similar phenomena would occur with other parylene family polymers, parylene-C for example. TDS also revealed that the first sign of thermal degradation begins at slightly below $400^{o}C$ with the appearance of a peak at 91 amu that is a product of the decomposition reaction. Thickness change due to isothermal annealing in vacuum revealed that thickness loss was not detectable after 120 minutes at $400^{o}C$ but was detectable after 120 minutes at $410^{o}C$. These results agree with those of other researchers.

7.   UV Degradation

It is commonly known that radiation from the sun and man-made sources can cause deteriorative ageing and degradation of polymers. The energy of a 250 nm photon is 480 kJ/mol and the strengths of C-C and C-H bonds are approximately 420 and 340 kJ/mol, respectively, and may be much less in certain environments, particularly in the neighborhood of aromatic or un-saturated structures like parylene aromatic ring [73]. The absorption of light with energies in excess of bond energies can lead to bond breakage and the production of free radicals. Free radicals are very reactive chemically and their production in a polymer can lead to chemical reactions with gases that are present (especially oxygen), propagating chain scission, and cross-linking via chain reactions [138].

As stated in Chapter 1, parylene films have found and continue to find many applications. More recently, parylene films have been investigated for use as a low dielectric constant interlayer dielectric for advanced very-large-scale-integration (VLSI) interconnect technologies [5, 35, 48, 136]. Future VLSI processing technologies are expected to include deep UV and vacuum UV ($<$ 200 nm) photolithography [54, 55, 137]. Potential interactions between materials used in VLSI manufacturing and the lithographic radiation need to be considered. Parylene is also used and has been researched for several space applications [27, 28] where an ample amount of high energy, low wavelength radiation exists.

The UV absorbance spectrum of parylene has been fully characterized [1]. Parylene does not absorb visible light but does absorb at the shorter wavelength, higher energy side of the ultraviolet range. The absorptivity increases about 300 times going from 300 nm to 200 nm, which is expected for the electronic system of the parylene benzene ring. Parylene has been shown to be patternable by laser ablation at a wavelength of 266 nm [56]. Recently, researchers have shown that photo-oxidation of parylene and a chlorinated version readily occurs at wavelengths $>$ 300 nm [57, 58].

The effect of ultraviolet (UV) radiation of $\lambda > 250$ nm on the thermal stability, electrical, and optical properties of thin parylene-N films has been studied [59, 150]. Evidence of slight oxidation of the UV treated film was seen using fourier-transform infrared spectroscopy. Thermal desorption spectrometry and thickness change after annealing were used to analyze the thermal stability of as-deposited and UV-treated parylene thin films. The thermal stability of the UV treated films was seen to decease as the radiation dose increased. Electrical measurements revealed an increase in the leakage current density, the dielectric constant, and the dissipation factor of UV treated films. No change was seen in the refractive index at 634 nm.

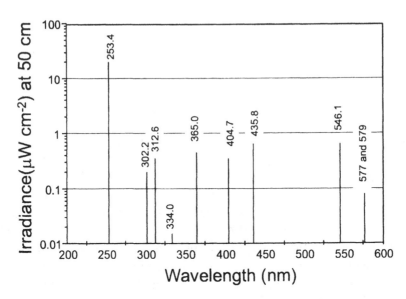

Figure 6.11: The irradiance of the 6035 Hg(Ar) lamp at 50 cm.

The output of the source lamp according for this research is shown in figure 6.11. The calculated dose for each wavelength for a 7 hour and a 24 hour exposure is shown in table 6.3. The dose for some of the same wavelengths that are output by the lamp is given for the sun outside of the earth's atmosphere as a reference [139].

A clear difference in the thermal stability of the treated vs untreated films can be seen from the thickness change as a function of annealing temperature data. The samples in the study were annealed for 2 hours at various temperatures from 300°C to 430°C under vacuum. The thickness change data for the as-deposited and UV treated samples is shown in figure 6.12. The initial film thickness for these samples was 3300 Å. The thermal stability of the film is degraded by the UV treatment. Film thickness decreases at temperatures as low as 300°C with significant degradation of the film occurring at 400°C. The control does not begin to have decreased thickness until 410°C, which is the degradation temperature of parylene-N.

The films receiving UV treatment were shown to have a much higher leakage current than as-deposited films (see figure 6.13). This figure shows leakage current density versus field strength for two samples from each group. As can be seen the leakage current density for the 24 hour UV treated samples has a region where the leakage values jump up and become erratic. Three out of the four capacitors that exhibited electrical breakdown (failure) after the 24 hour UV treatment failed in this region. Those that did not breakdown appeared to be healed. It was found that

Table 6.3: Calculated irradiance and dose of the Hg(Ar) arc
lamp for 7 and 24 hour exposures. Irradiance from
the sun outside the earth's atmosphere is given as a
reference.

Wavelength (nm)	Irradiance from Hg(Ar) at 10 cm ($\mu W cm^{-2} nm^{-1}$)	Irradiance from sun outside atmosphere ($\mu W cm^{-2} nm^{-1}$)	Dose from lamp for 7 hr exposure ($J cm^{-2}$)	Dose from lamp for 24 hr exposure ($J cm^{-2}$)
253.7	500	5.254	12.6	43.2
302.2	5	44.86	0.126	0.432
312.6	10	70.32	0.253	0.864
334.0	0.4	97.10	0.009	0.032
365.0	11	..	0.284	0.972
404.7	9	---.	0.221	0.757
435.8	16	—	0.410	1.404
546.1	16	.-..	0.410	1.404
557	2	. ..	0.050	0.173
559	2	.- .	0.050	0.173

[a] [139]

the top Al electrode of these what look to be healed capacitors had a
number of holes after testing. Therefore there was probably no healing
taking place, just an evaporation or blow-out of the electrode due to
heat generation in the high leakage area removing it from the circuit.
Table 6.4 shows the number of capacitors exhibiting dielectric breakdown
from each test group and the field strength at which they occurred. The
inclusion of oxygen and of new double bonds in the polymer chain can
lead to increased conduction along a molecule due to the unsaturated
structure. Also, broken bonds, which can be satisfied by the removal or
donation of an electron, or both, behave as states within the band gap.
Although these states are localized around the area of the defect and
electrons or holes entering the traps are not available for conduction,
conduction processes are possible if the density of the defects or the
applied field is high enough [138].

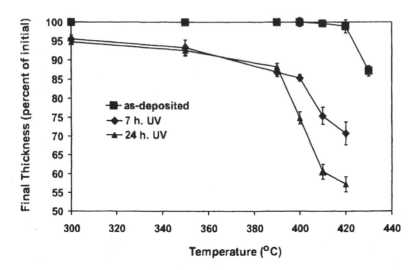

Figure 6.12: Film Thickness change due to degradation from annealing for 2 hours under vacuum. Data is shown for as-deposited, 7 hour, and 24 hour UV-treated samples.

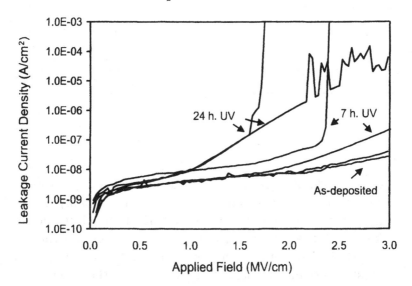

Figure 6.13: Direct current leakage density versus field strength for as-deposited, 7 hour, and 24 hour UV-treated samples.

The dielectric constant and dissipation factor for films receiving UV treatment were increased as the dose increased, see table 6.5. These increases can also be caused by the inclusion of double or broken bonds

Table 6.4: Capacitors exhibiting dielectric breakdown and the associated field strength for as-deposited, 7 hr. UV, and 24 hr. UV-treated samples.

Sample	Capacitors exhibiting breakdown	Field at which breakdown occurred (MV/cm)
As-Deposited	1/15	2.35
7 hrs. UV-Treatment	1/15	2.11
24 hrs. UV-Treatment	4/15	0.06,1.99,2.35,2.86

Table 6.5: Dielectric constant and dissipation factor values for as-deposited, 7 hr. UV, and 24 hr. UV-treated samples.

Sample	Dielectric Constant	Dissipation Factor
As-Deposited	2.61 ± 0.04	0.006 ± 0.001
7 hrs. UV-Treatment	2.94 ± 0.02	0.015 ± 0.001
24 hrs. UV-Treatment	3.11 ± 0.01	0.021 ± 0.001

into the structure as de-localized electrons are polarizable. Also, the inclusion of oxygen leads to an increase in polarization, which will increase the dielectric constant.

It is evident from this study that UV radiation with $\lambda > 250$ nm can deteriorate the thermal stability and electrical properties of parylene thin films. The minimum dose of 248 nm radiation used in this study, 12 J cm^{-2}, caused measurable degradation of the film's properties. Deep UV lithography uses typical doses of 0.03 to 0.3 J cm^{-2} [140, 54, 141], so the degradation of parylene due to interaction with the lithographic procedure is probably not likely, especially since the intensity of the radiation would be attenuated by the photoresist layer. On the other hand, an equivalent of 12 J cm^{-2} of 254 nm energy is radiated from the sun outside the earth's atmosphere in about 700 hours, or 30 days. This is potentially harmful for parylene used in space applications where the film will have exposure to such radiation. Also, other potential applications of parylene in which the film will have exposure to UV light from a man-made source need to be evaluated for their effects on the film's properties pertinent to that application.

Chapter 7

OTHER CVD POLYMERS

1.   Introduction

In addition to the Parylene family, there are several classes of polymer that have been shown to form thin film by evaporation or via chemical vapor deposition routes. Among them, rigid-rod aromatic polymers, due to their structure, possess several very attractive properties that have a great potential for many applications such as protective coating. They possess high thermal stability, high glass transition temperature, high hardness, insolubility to common solvents, and oxidation resistance. Although these films possess very desirable properties, only limited work has been reported in this area of research. In this Chapter, we will describe a selective group of polymeric films deposited via different types of vapor deposition techniques.

2.   Polynaphthalene

Poly(1,4-naphthalene), referred to polynaphthalene, as shown in Figure 7.1, can be synthesized in the form of powder using solution methods [151]. The powder-like material has low molecular weight and is easily soluble. Polynaphthalene can be electrochemically deposited to form films [152]. CVD of high molecular weight insoluble polynaphthalene films is also possible [153, 154]. The CVD of polynaphthalene is solvent and catalyst free, similar to parylene in this manner.

2.1   Film deposition

The precursor for depositing polynaphthalene via CVD is the liquid form *o*-diethynylbenzene. The precursor is evaporated at room temperature or below and is fed into a vacuum deposition system. The precursor

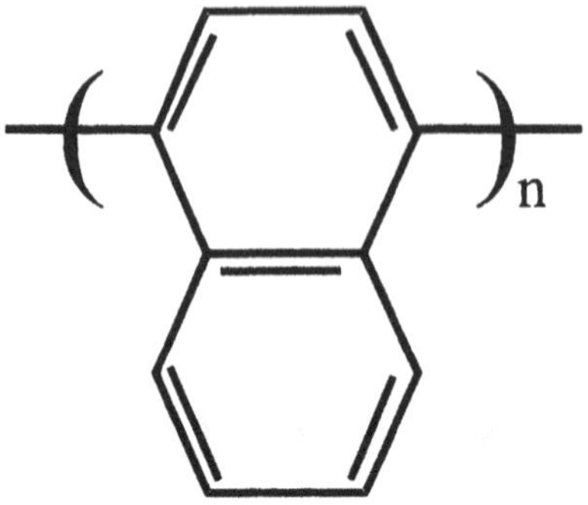

Figure 7.1: The chemical structure of poly(1,4-naphthalene).

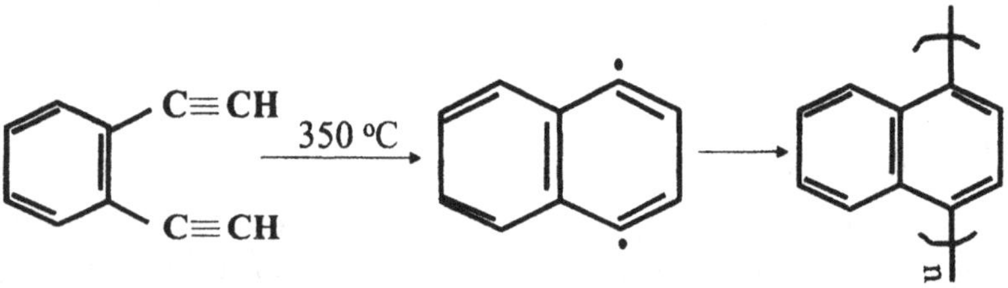

Figure 7.2: The precursor o-diethynylbenzne vapor is heated in a CVD reactor at above 350°C to form poly-naphthalene via a diradical route.

vapor can be thermally activated to form diradical monomer at about 350°C. Unlike parylene CVD, in the experiments reported so far, the monomer synthesis step for polynaphthalene is not spatially uncoupled from the polymer deposition step. This means that the film deposition occurs at the same time as monomer formation. The chemical route for the polymer film formation is shown in figure 7.2.

Both cold-wall and hot-wall reactors can be employed to deposit poly-naphthalene films. Figures 7.3 and 7.4 are schematic diagrams showing the reactor design for cold-wall and hot-wall reactors, respectively. In the cold-wall reactor, the precursor vapor evaporated from the source enters the chamber with a base pressure on the order of mTorr. After entering the chamber the vapor pass through a diffuser before it reaches the substrate. In this deposition scheme, one can prevent the deposition and reaction on the chamber walls. The diradicals only can react and form a polymeric film on the heated substrate (350-400°C). The deposition pressure is usually less than 1 Torr. However, it had been shown that the deposition rate is rather low using a cold wall reactor. This design is not suitable if one wants to grow films on the order of microns.

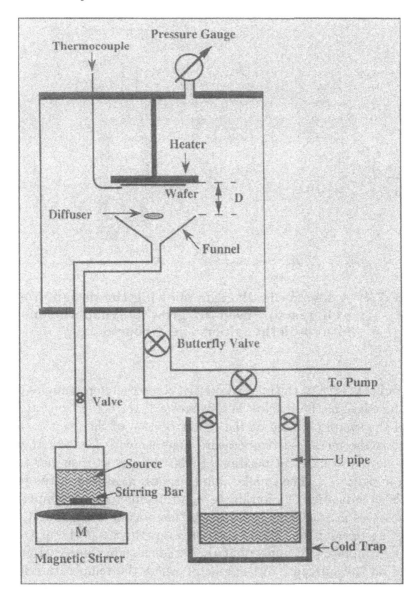

Figure 7.3: A schematic diagram showing the design of a cold-wall reactor used to grow polynaphthalene film from *o*-diethynylbenzene precursor.

In the hot-wall reactor, the precursor vapor is directed into a reactor through a needle valve as shown in figure 7.4. The substrate temperature is around 350-400°C. The pressure ranges between few hundreds mTorr to a few Torr. The deposition rate ranges from 1 to 10 nm/s. Tempera-

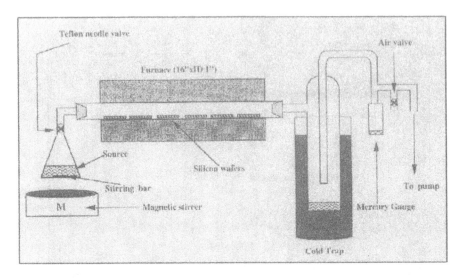

Figure 7.4: A schematic diagram showing the design of a hot-wall reactor used to grow polynaphthalene film from o-diethynylbenzene precursor.

ture gradients exist at the inlet and outlet zones therefore samples placed near the edges of the reactor would have a lower deposition rate than that of the samples placed at the center portion of the reactor. Figure 7.5 shows the measured temperature together with the film thickness data collected at different position of a 30" long reactor similar to that shown in figure 7.4. The smaller film thickness measured at the reactor edges is an indication of lower deposition rate due to lower temperature. One observes a reasonably uniform thickness along a 15" length. It is interesting to note that the deposition rate reduces at the exit of the reactor. It appears to imply that the diradicals are consumed and deposited on the substrate in the reactor where the temperature is high and there is no remaining diradicals left when the vapor emerges from the reactor. Another possibility is that there are still abundant of diradicals at the exit of the reactor but they do not undergo polymerization at lower temperature. In any case, the substrate has to be maintained at an elevated temperature to achieve polynaphthalene deposition in these types of reactor design.

   Fluorinated polynaphthalene can also be vapor deposited through a CVD route using a precursor of 1,2-diethynyltetrafluorobenzene. The reaction to form fluorinate polynaphthalene through the diradical route is shown in figure 7.6. At similar deposition conditions, the deposition rate has been observed to be higher than that of the polynaphthalene.

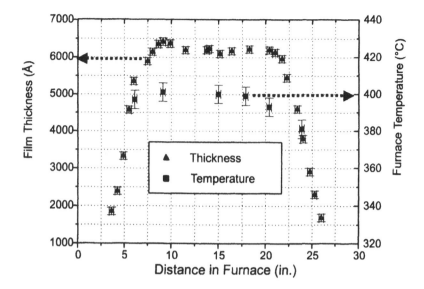

Figure 7.5: This figure shows the measured temperature along the furnace plotted with the film thickness data collected a different positions along a 30 inch section of a hot wall deposition chamber.

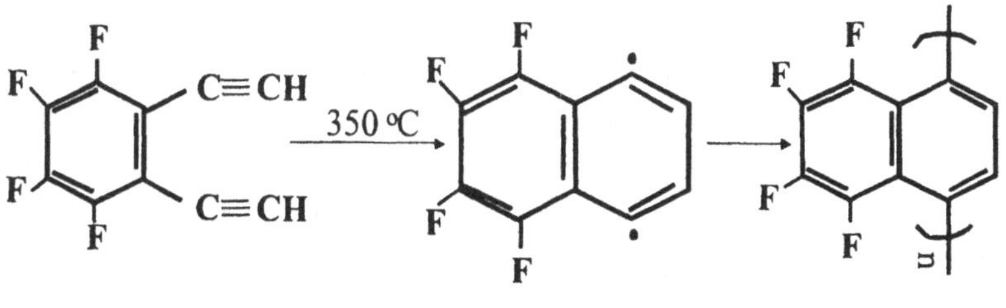

Figure 7.6: Synthesis of fluorinated polynaphthalene from 1,2-diethynyltetrafluorobenzene.

2.2 Polynaphthalene Film Properties

Polynaphthalene film adheres well to dielectric surfaces such as glass or Si surface and is not soluble in common solvents. Polynaphthalene film hardness is 2 GPa which is roughly one order of magnitude harder than that of the Parylene film. Polynaphthalene films possess exceptionally high thermal stability. Polynaphthalene can stand a temperature of 500°C without developing cracks or thickness change. However oxygen incorporation in the films can adversely affect the thermal stability and electrical properties of the films. Oxygen reacts strongly at the surface

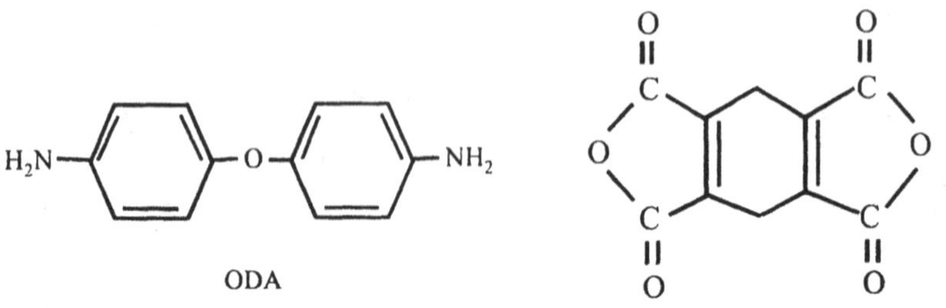

Figure 7.7: Example of a diamine (ODA) and a dianhydride (PMDA) monomer.

during deposition and can be easily incorporated into the film during deposition if it is present. It would be therefore very desirable to deposit the film in a system with a good base pressure and low leakage to maintain a low level of oxygen contamination. Overall, the mechanical and thermal properties of polynaphthalene are very attractive for many practical coating applications. A major challenge is the high deposition temperature that is required to form the film.

3.   Polyimides

Polyimides are one of the most commonly used polymers in electronics industry. They are normally deposited through wet chemistry such as spin-coated and cured. Chemical vapor deposition of polyimides under a vacuum condition (dry) is also possible [155]. This is achieved through a co-polymerization process of a diamine monomer with a dianhydride monomer by a CVD. In this process, the aromatic diamine and dianhydride are evaporated in a vacuum chamber and the evaporation rate is controlled separately for the two components. The co-evaporated film with the stoichiometric ratio is in a form of polymeric acid. The film is then heated to an elevated temperature for imidization to form the polyimide film. Typically the imidization temperature is around 200°C. The rod-like polymide films have higher temperature stability than polymers of the parylene family. They dissociate at above 450°C.

Examples of diamine are O-tolidine(OTD), 3,3'-dimethyl 4,4-diaminodiphenyl methene (MeMDA), 2,2-bis[4-(4-amino-phenoxy)phenyl]-hexafluoropropane (BDAF), 2,2-bis[4-(4-amino-phenoxy)phenyl]propane, and 4,4'-oxydianiline (ODA). Examples of dianhydride are pyromellitic dianhydride (PMDA), 3,3'-4,4'-biphenyl tetracar-boxylic dianhydride , and 4,4'-(hexafluoroisopropylidene) bis (phthalic anhydride). Figure 7.7 displays the chemical structure of one diamine monomer and one dianhydride monomer.

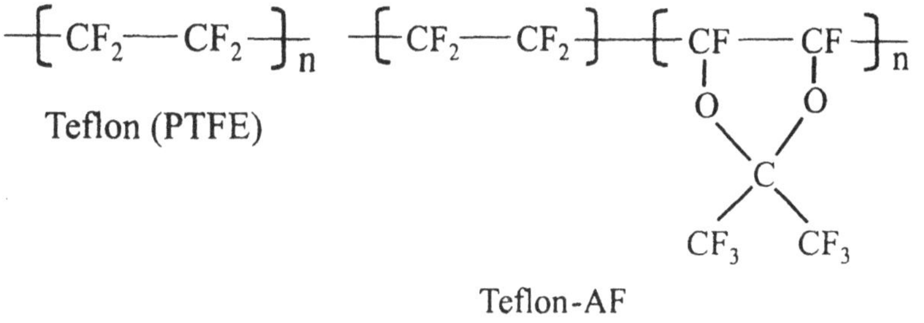

$$-\left[CF_2-CF_2\right]_n \quad -\left[CF_2-CF_2\right]-\left[\begin{array}{cc} CF & - & CF \\ | & & | \\ O & & O \end{array}\right]_n$$

Teflon (PTFE)

$$\begin{array}{c} O \quad O \\ \diagdown \quad \diagup \\ C \\ \diagup \quad \diagdown \\ CF_3 \quad CF_3 \end{array}$$

Teflon-AF

Figure 7.8: Chemical structure of Teflon and Teflon-AF.

4.   Polytetrafluoroethylene Family

Polytetrafluoroethylene (or PTFE) also trademarked as Teflon by DuPont is a well-known low index of refraction and low dielectric constant polymeric material. Thin film Teflon can be produced by the pyrolysis of bulk Teflon in vacuum [156]. The vapor, when condensed on a substrate is repolymerized. Figure 7.8(left) shows the chemical structure of Teflon. Teflon films typically possess a high degree of crystallinity. An interesting derivative of Teflon is Teflon-AF. Teflon-AF is comprised of a number of Teflon-like copolymers which incorporate chemical groups to inhibit cystallization. These amorphous Teflon-AF materials include a Teflon-AF 1600, a copolymer of tetrafluoroethylene and 66 mol percent of 2,2-bis(trifluoromethyl)-4,5-difluoro-1,3-dioxole. The chemical structure is shown in figure 7.8 (right). Because of the lack of crystallinity, Teflon-AF is optically transparent. Also, the dielectric constant is very low at 1.9 and index of refraction is below 1.3.

Teflon-AF can also be repolymerized by the pyrolysis of bulk Teflon-AF [157]. It can be deposited in a base pressure of $10^{(-6)}$ Torr by heating the source material above 400°C in a graphite crucible. The vapor contains volatile fragments of Teflon-AF and condenses onto a substrate at room temperature or below to form the Telfon-AF film. In order to control the deposition rate, it is desirable to use a crucible with a small nozzle. There is evidence that the source material actually melts during deposition. Detailed study using FTIR (Fourier transform infrared spectroscopy) indicates that the film structure and composition are identical compared to that of the source material. It is quite remarkable that a copolymer of this complexity can be repolymerized at the substrate after the fragmentation of the source material during the evaporation process.

References

[1] W. Beach, C. Lee, and D. Bassett, Encyclopedia of Polymer Science and Engineering (Wiley, New York, 1985), 17, 990.

[2] J.W. Nicholson, *The Chemistry of Polymers* (The Royal Society of Chemistry, Cambridge, 1991), 35.

[3] R.B. Seymour and C.E. Carraher, *Polymer Chemistry*(Marcel Dekker, Inc., New York, 1988), 332.

[4] J.R. Fried, *Polymer Science and Technology* (Prentice Hall, New Jersey, 1995), 49.

[5] T.-M. Lu and J.A. Moore, MRS Bulletin 22(10), 28(1997).

[6] R. Sharangpani and R. Singh, Mat. Res. Soc. Symp. Proc. V476, 207(1997).

[7] S. Ganguli, Ph.D. Thesis, Rensselaer Polytechnic Institute, Troy, NY, 1997.

[8] R. d'Agostino, *Plasma Deposition, Treatment, and Etching of Polymers*, (Academic Press, New York, 1990) Chpt. 1.

[9] W.W. Lee, R. d'Agostino, and M.R. Wertheimer, Mat. Res. Soc. Syp. Proc. V544, (1999).

[10] M. Szwarc, Discussions Faraday Soc. 2, 46(1947).

[11] M. Szwarc, J. Polym. Sci. 6(3), 319(1951).

[12] M. Szwarc, J. Chem. Phys. 16(2), 128(1948).

[13] W. Gorham, J. Polym. Sci. Part A-1, 4, 3027(1966).

[14] A. Greiner, *Polymeric Materials Encyclopedia Vol.9* (CRC Press, New York,1996), 7171.

[15] D. Martini, K. Shepherd, R. Sutcliffe, J. Kelber, H. Edwards, and R. San Martin, Appl. Sur. Sci. 141, 89(1999).

[16] A.J. Roche and W.R.D Olbier Jr., J. Org. Chem. 64, 9137(1999).

[17] M. Morgen, S.-H. Rhee, J.-H. Zhao, I. Malik, R. Ryan, H.-H. Ho, M.A. Plano, and P. Ho, Macromolecules 32, 7555(1999).

[18] J.J. Senkevich, V. Simkovic, and S.B. Desu, Mat. Res. Soc. Symp. Proc. V511, 139(1998).

[19] D. Mathur, G.-R. Yang, and T.-M. Lu, J. Mater. Res. 14(1), 246(1999).

[20] L.A. Errede, R.S. Gregovian, and J.M. Hoyt, J. Am. Chem. Soc. 82, 436(1960).

[21] W.F. Gorham, Adv. in Chem. Series 91, 641(1969).

[22] J.F. Gaynor, S.B. Desu, and J.J. Senkevich, Macromolecules 28, 7343(1995).

[23] J.F. Gaynor, S.B. Desu, and J.J. Senkevich, J. Mater. Res. 11(7), 1842(1996).

[24] K.J. Taylor, M. Eissa, J.F. Gaynor, S.-P. Jeng, and H. Nguyen, 1997 Spring MRS Symposium.

[25] L. Alexandrova and R. Vera-Graziano, *Polymeric Materials Encyclopedia,* (CRC Press, New York, 1996), 7180.

[26] H.A. Carter, J. Chem. Ed. 73(12), 1160(1996).

[27] B.Q. Lee, R.H. Maurer, E. Nhan, and A. Lew, Int. J. of Microcircuits Electron. Packag. 22, 104(1999).

[28] M. Billot and C. Val, Proc. SPIE- Int. Soc. Opt. Eng. V3830, 74(1999).

[29] S. Petrovic, C. Brown, A. Ramirez, B. King, T. Maudie, D. Stanerson, G. Bitko, J. Matkin, J. Wertz, and D.J. Monk, Advances in Electronic Packaging 1997, INTERpack 97, 1, 2(1997).

[30] R. De Reus, C. Christensen, S. Weichel, S. Bouwstra, J. Lanting, G.F. Eriksen, K. Dyrbye, T.R. Brown, J.P. Krog, O.S. Jensen, and P. Gravesen, Microelectron. Reliab. 38(6-8), 1251(1998).

[31] X. Tang, J.M. Yang, Y.-C. Tai, and C.-M. Ho, Sensors and Actuators A73(1-2), 184(1998).

[32] K. Minami, H. Morishita, and M. Esaski, Sensors and Actuators A72(3), 269(1999).

[33] H. Hyun and P. Jae-Kyung, J. of Korea Inst. Tele. Elec. 34D(8), 80(1997).

[34] K.H. Stephan, H. Braeuninger, C. Reppin, H. Maier, D. Frischke, M. Krumrey, P. Mueller, Proc. SPIE-Int. Soc. Opt. Eng. 1743, 192(1992).

[35] H. Tetsuga, Mater. Sci. Eng. R23, 6, 243(1998).

[36] S. Ganguli, H. Agrawal, B. Wang, J. Mcdonald, T.-M. Lu, G.-R. Yang, and W. Gill, J. Vac. Sci. Technol. A 15(6), 3138(1997).

[37] G.-R. Yang, S. Ganguli, J. Karcz, W. Gill, and T.-M. Lu, J. Crystal Growth 183(3), 385(1998).

[38] D. Price, R. Gutmann, and S. Murarka, Thin Solid Films 308, 523(1997).

[39] K. Gilleo and G. Bacon, Vacuum Technology and Coating 9, 60(2000).

[40] G.E. Loeb, R.A. Peck, and J. Martyniuk, J. Neurosci. Methods 63(1-2), 175(1995).

[41] E.M. Schmidt, M.J. Bak, and P. Christensen, J. Neurosci. Methods 62(1-2), 89(1995).

[42] A.B. Fontaine, K. Koelling, S.D. Passos, J. Cearlock, R. Hoffman, and D.G. Spigos, J. Endovasc. Surg. 3(3), 276(1996).

[43] J.H. Noar, A. Wahab, R.D. Evans, and A.G. Wojcik, Euro. J. Ortho. 21(6), 685(1999).

[44] NIST Mass Spec Data Center, S.E. Stein, Director, "IR and Mass Spectra" in NIST Chemistry WebBook, NIST Standard Reference Database Number 69, Eds. W.G. Mallard and P.J. Linstrom, Nov 1998, National Institute of Standards and Technology, Gaithersburg MD, 20899 (http://webbook.nist.gov)

[45] J.B. Fortin and T.-M. Lu, J. Vac. Sci. Technol. A 18(5), 2459(2000).

[46] W.F. Beach, Macromolecules 11(1), 72(1978).

[47] J.F. Gaynor, Electrochemical Society Proc. V97-8, 176(1997).

[48] S. Rogojevic, J.A. Moore, and W.N. Gill, J. Vac. Sci. Technol. A 17(1), 266(1999).

[49] B. O'Shaughnessy and U. Sawhney, P. R. L. 76(18), 3444(1996).

[50] J. Charlier, C. Detalle, F. Valin, C. Bureau, and G. Lecayon, J. Vac. Sci. Technol. A 15(2), 353(1997).

[51] R. Biscarini, P. Samori, O. Greco, and R. Zamboni, P. R. L. 78(12), 2389(1997).

[52] F. Cacialli and P. Bruschi, J. Appl. Phys. 80(1), 70(1996).

[53] G.W. Collins, S.A. Letts, E.M. Fearon, R.L. McEachern, and T.P. Bernat, P. R. L. 73(5), 708(1994).

[54] Q.Y. Lin, A. Cheng, W.W. Ma, and J. Cheng, Proc. SPIE - Int. Soc. Opt. Eng. Proc. SPIE - Int. Soc. Opt. Eng. V3679, 942(1999).

[55] K. Matsumoto and K. Suwa, AIP Conference Proceedings V449, 484(1998).

[56] J.D. Weiland, D.J. Anderson, C.C. Pogatchnik, and J.J.B oogaard, Proc. of the 19th Annual Int. Conf. IEEE Eng. Med. Bio. V5 2273(1997).

[57] M. Bera, A. Rivaton, C. Gandon, and J.L. Gardette, European Polymer Journal 36, 1753(2000).

[58] M. Bera, A. Rivaton, C. Gandon, and J.L. Gardette, European Polymer Journal 36, 1765(2000).

[59] J.B. Fortin and T.-M. Lu, Thin Solid Films 397, 223(2001).

[60] T.E. Nowlin and D.F. Smith, J. Electrochem. Soc. 143(6), 1619(1980).

[61] J.T.C. Yeh and K.R. Grebe, J. Vac. Sci. Technol. A1(2), 604(1983).

[62] R.D. Tacito and C. Steinbruchel, J. Electrochem. Soc. 143(6), 1974(1996).

[63] B. Ratier, Y.S. Jeong, A. Moliton, and P. Audebert, Optical Materials 12, 229(1999).

[64] H. Yasuda, J. Macromol. Sci. Chem. A10(3), 383(1976).

[65] E.M. Liston, J. Adhesion, 30, 199(1989).

[66] L. Trazbon and O.O. Awadelkarim, Semicond. Sci. Technol. 15, 309(2000).

[67] J.J. Senkevich, C.J. Mitchell, A. Vijayaraghaven, E.V. Barnat, J.F. McDonald, and T.-M. Lu, J. Vac. Sci. Technol. A 20(4), 1445(2002).

[68] J. Lahann, M. Balcells, T. Rodon, J. Lee, I.S. Choi, K.F. Jensen, and Robert Langer, Langmuir 18, 3632(2002).

[69] J. Lahann and R. Langer, Macromolecules 35, 4380(2002).

[70] J. Lahann, H. Hocker, and R. Langer, Angew. Chem. Int. Ed 40(4), 726(2001).

[71] J. Lahann and R. Langer, Macromol. Rapid Commun. 22, 968(2001).

[72] G.H. Meeten, *Optical Properties of Polymers* (Elsevier Applied Science Publishers, New York, 1986) Chpt. 1.

[73] N. Grassie and G. Scott, *Polymer Degradation and Stabilization* (Cambridge Univ. Press, New York, New York, 1985) Chpt.1-3.

[74] A. Chowdhury, W. Read, G. Rubloff, L. Tedder, and G. Parsons, J. Vac. Sci. Technol. B 15(1), 127(1997).

[75] K. Okimura and N. Maeda, J. Vac. Sci. Technol. A 16(6), 3157(1998).

[76] C. Courtney, B. Smith, and H. Lamb, J. Electrochem. Soc. 145(11), 3957(1998).

[77] J.T.S. Andrews and E.F. Westrum, Jr. J. Phys. Chem. 74(10), 2170(1970).

[78] R.H. Boyd, Tetrahedron 22, 119(1966).

[79] K. Nishiyama, M. Sakiyama, S. Seki, H. Horita, T. Otsubo, and S. Seki, Tetrahedron Letters 42, 3739(1977).

[80] R.P. Mariella, S.W. Steinhauser, and A.C. Diebold, J. Vac. Sci. Technol. B 5(5), 1360(1987).

[81] S. Shimizu, N. Nakashima, and Y. Sakata, Chem. Phys. Let. 284, 396(1998).

[82] L.A. Errede and J.M. Hoyt, Quarterly Reviews V82, 436(1960).

[83] L.A. Errede and M. Szwarc, Quarterly Reviews, 301(1957).

[84] O. Schäfer, F. Brink-Spalink, and A. Greiner, Macromol. Rapid Commun. 20, 190(1999).

[85] L.A. Auspos, L.A.R. Hall, J.K. Hubbard, W.M. Kirk, J.R. Schaefgen, and S.B. Speck, J. Poly. Sci. 15, 9(1955).

[86] M. Hamming and N. Foster, *Interpretation of Mass Spectra of Organic Compounds* (Academic Press, NY, 1972), chapter 5.

[87] M. Rao and C. Dong, J. Vac. Sci. Technol. A 15(3), 1312(1997).

[88] R. Johnstone, *Mass Spectrometry for Organic Chemists* (Cambridge University Press, London, 1972), 60.

[89] T. Koday and M. Hampden-Smith, *The Chemistry of Metal CVD* (Weinheim, New York, 1994), 469.

[90] C.E. Morosanu, *Thin Films by Chemical Vapor Deposition* (Elsevier, New York, 1990), Chpt. 5.

[91] M.L. Hitchman and K.F. Jensen, *Chemical Vapor Deposition* (Academic Press Inc., San Diego, CA, 1993), 39.

[92] C.H.L. Goodman, *Crystal Growth, Theory and Techniques* (Plenum Press, New York, 1980), 21.

[93] E.M. Charlson, E.J. Charlson, and R. Sabeti, IEEE Transactions on Biomedical Engineering 33(2), 202(1992).

[94] F.E. Cariou, D.J. Vally, and W.E. Loeb, IEEE Trans. Parts, Mater. Pacakg., 54(1965).

[95] P. Kramer, A.K. Sharma, E.E. Hennecke, and H. Yasuda, Journal of Polymer Science 22, 475(1984).

[96] C.N. Satterfield, *Heterogeneous Catalysis in Practice* (McGraw-Hill, New York, 1980), 36.

[97] X. Zhang, S. Dabral, C. Chiang, J.F. McDonald, and B. Wang, Thin Solid Films 270, 508(1995).

[98] J.B. Fortin and T.-M. Lu, Chem. Mater. 14, 1945(2002).

[99] P.J. Flory, *Principles of Polymer Chemistry* (Cornell Univ. Press, Ithaca, 1953), 514.

[100] C. Aharoni and F.C. Tompkins, Advances in Catalysis 21, 1(1970).

[101] J.F. O'Hanlon, *A User's Guide to Vacuum Technology, 2nd Ed.* (Wiley, New York, 1989), Chpt.3.

[102] S. Kubo and B. Wunderlich, J.Poly.Sci., Polym. Phys. Ed. 10, 1949(1972).

[103] D.O. Hayward and B.M.W. Trapnell, *Chemisorption* (Butterworth and Co., London, 1964), Chpt. 3.

[104] A. Zangwill, *The Physics at Surfaces* (Cambridge University Press, Cambridge, 1988), Chpt.9,14.

[105] P. Kisliuk, J. Phys. Chem. Solids 3, 95(1957).

[106] K.M. Vaeth and K.F. Jensen, Chem. Mater. 12, 1305(2000).

[107] S. Brunauer, *The Adsorption of Gases and Vapors* (Princeton Univ. Press, London, 1943), Chpt 8.

[108] V. Lachet, A. Boutin, B. Tavitian, and A.H. Fuchs, Langmuir 15, 8678(1999).

[109] Y.-P. Zhao, J.B. Fortin, G. Bonvallet, G.-C. Wang, and T.-M. Lu, Phy. Rev. Let. 85(15), 3229(2000).

[110] F. Family and T. Viscsek, Editors, *Dynamics of Fractal Surfaces* (World Scientific, Singapore, 1991).

[111] A.-L. Barabási and H.E. Stanlye, *Fractal Concepts in Surface Growth* (Cambridge University Press, Cambridge, England, 1995).

[112] P. Meakin, *Fractals, Scaling, and Growth far from Equilibrium* (Cambridge University Press,Cambridge, England, 1998).

[113] C.P. Wong, Editor, *Polymers for Electronic and Photonic Applications* (Academic Press, Boston, 1993).

[114] T.-M. Lu, H.-N. Yang, and G.-C. Wang in *Fractals Aspects of Materials*, edited by F. Family, P. Meakin, B. Sapoval, and R. Wool, MRS Symposia Proceedings No 367 (Materials Research Society, Pittsburgh, 1995), p. 283.

[115] J. Aué and J.Th.M. De Hosson, Appl. Phys. Lett. 71, 1347(1997).

[116] Z.W. Lai and S. Das Sarma, Phys. Rev. Lett. 66, 2349(1991).

[117] P.K. Wu, G.-R. Yang, J.F. McDonald, and T.-M. Lu, J. Electronic Mat. 24(1), 53(1995).

[118] B.L. Joesten, J. Appl. Poly. Sci. 18, 439(1974).

[119] H.H.G. Jellinek and S.N. Lipovac, J. Poly. Sci. A-1 8, 2517(1970).

[120] H.H.G. Jellinek and S.H. Ronel, J. Poly. Sci. A-1 9, 2605(1971).

[121] S.L. Madorsky and S. Straus, J. Research of the National Bureau of Standards 55(4), 223(1955).

[122] S.C. Selbrede and M.L. Zucher, Mat. Res. Soc. Symp. Proc. 476, 219(1997).

[123] S.Y. Lazareva, A.V. Osipov, and Y.E. Malkov, Vysokomol. Soyed. A21(7), 1509(1979).

[124] C. Chiang, A.S. Mack, C. Pan, Y.-L. Ling, and D.B. Fraser, Mat. Res. Soc. Symp. Proc. 381, 123(1995).

[125] X. Zhang, S. Dabral, C. Chiang, J.F. McDonald, and B. Wang, Thin Solid Films 270, 508(1995).

[126] D.J. Monk, H.S. Toh, and J. Wertz, Sensors and Materials 9(5), 307(1997).

[127] T.E. Nowlin, D.F. Smith, and G.S. Cieloszyk, J. Poly. Sci. Chem. Ed. 18, 2103(1980).

[128] T.E. Baker, G.L. Fix, and J.S. Judge, J. Electrochem. Soc. Solid State Science and Technology 127(8), 1851(1980).

[129] J.M. Shaw, M.A. Frish, and F.H. Dill, IBM J. Res. Develop. 5, 219(1977).

[130] M.E. Fragala, G. Compagnini, L. Torrisi, and O. Puglisi, Nucl. Instr. Meth. Phys. Res. B 141, 169(1998).

[131] M. Sato, M. Iijima, and Y. Takahashi, Thin Solid Films 308-309, 90(1997).

[132] A.Vanbennokom, P. Willemsen, and R.J. Gaymans, Polymer 37, 5447(1996).

[133] C. Chang, A.S. Mack, C. Pan, Y.-L. Ling, and D.B. Fraser, Mat. Res. Soc. Symp. Proc. V381, 123(1995).

[134] J.J. Senkevich, J. Vac. Sci. Technol. A 18(5), 2586(2000).

[135] C.G. Kullingel, M. Tomozawa, and S.P. Murarka, J. Electrochem. Soc. 145(5), 1790(1998).

[136] S. Ganguli, H. Agrawal, B. Wang, J. Mcdonald, T.-M. Lu, G.-R. Yang, and W. Gill, J. Vac. Sci. Technol. A 15, 3138(1997).

[137] R. Mih, N. Chen, K. Jatzen, J. Marsh, and S. Schneider, Proc. SPIE - Int. Soc. Opt. Eng. V3679, 827(1999).

[138] L.A. Dissado, C. Fothergill, Electrical Degradation and Breakdown in Polymers, (Peter Peregrinus Ltd., London, UK, 1992), Chapters 2,3,9.

[139] T. Gerard, M. Herse, P.C. Simon, D. Labs, H. Mandel, D.Gillotay, Solar Physics 171, 283(1997).

[140] J.P. Kuijten, T. Harris, and L. vanderHeijden, Proc. SPIE- Int. Soc. Opt. Eng. V4000, 843(2000).

[141] G.R. Misium, M. Tipton, and C.M. Garza, J. Vac. Sci. Technol. B 8, 1749(1990).

[142] S. Wolf and R.N. Tauber, Silicon Processing for the VLSI Era, Volume 1 - Process Technology, (Lattice Press, Sunset Beach, CA, 1986).

[143] A.C. Fozza, A. Bergeron, J.E. Klemberg-Sapieha, and M.R. Wertheimer, Mat. Res. Soc. Symp. Proc. V544, 109(1999).

[144] R. Wilken, A. Hollander, and J. Behnisch, Mat. Res. Soc. Symp. Proc. V544, 223(1999).

[145] R. Lamendola, E. Martarrese, M. Creatore, P. Favia, and R. d'Agostino, Mat. Res. Soc. Symp. Proc. V544, 269(1999).

[146] J.E. Klemberg-sapieha, L. Martinu, N.L.S. Yamasaki, and C.W. Lantman, Mat. Res. Soc. Symp. Proc. V544, 277(1999).

[147] M.S. Hargreaves, D.S. Hussey, and R.E. Leuchtner, Mat. Res. Soc. Symp. Proc. V544, 291(1999).

[148] D.T. Clark and A. Dilks, J.Poly.Sci. 15, 2321(1977).

[149] J.M. Blakely, *Introduction to the Properties of Crystal Surfaces*, (Pergamon Press, New York, 1973).

[150] J.B.Fortin, *Poly-para-xylylene Thin Films: A Study of the Deposition Chemistry, Kinetics, Film Properties, and Film Stability*, Ph.D. Thesis, Rensselaer Polytechnic Institite, 2001.

[151] J.G. Speight, P. Kovacic, and F.W. Koch, J. Macromol. Sci. Revs. Macromol. Chem., C5(2), 295(1971).

[152] S. Zecchin, R. Tomat, G. Schiavon, and G. Zotti, Synth. Met. 25, 393(1988).

[153] C.-I. Lang, G.-R. Yang, J.A. Moore, and T.-M. Lu, Mat. Res. Soc. Proc. 381, 45(1995).

[154] J.A. Moore, C.-I. Lang, T.-M. Lu, and G.-R. Yang, Polym. Mater. Sci. Eng. 72, 437(1995).

[155] T. Strunskus and M. Grunze, in *Polyimides*, edited by M. Ghosh and K. Mittal, Dekker, New York, 187(1996).

[156] W. de Wilde and G. de Mey, Vacuum 24, 307(1973).

[157] T.C. Nason, J.A. Moore, and T.-M. Lu, Appl. Phys. Lett. 60, 1866(1992).

[158] ASTM, Annual Book of Standards Vol 06.01, 1997.

[159] R.J. Clark, IBM Technical Disclosure Bulletin, 20(3), 945(1977).

[160] T. Riley, T. Cobo Mahuson, and K. Seibert, CECON 1980, pg 93.

[161] R.K. Sadhir, W. J. James, H.K. Yasuda, A.K. Sharma, M.F. Nichols, and A.W.Hahn, Biomaterials 2, 239(1981).

[162] A.K. Sharma and H. Yasuda, J. Vac. Sci. Technol. 21(4), 994(1982).

[163] G. Surendran, W.J. James, M. Gazicki, and H. Yasuda, J. Poly. Sci. A: Poly. Chem. 25, 2089(1987).

[164] J.J. Senkevich, S.B Desu, and V. Simkovic, Polymer 41, 2379(2000).

[165] C.J. Brown and A.C. Farthing, J. Chem. Soc., 3270(1953).

[166] R. Iwamoto and B. Wunderlich, J. Poly. Sci. Polymer Physics Edition 11, 2403 (1973)

[167] J. C. Salamone editor, Polymeric Materials Encyclopedia (CRC Press, New York, 1996), 7173.

[168] J.J. Senkevich and S.B. Desu, Polymer 40, 5751(1999).

[169] J.J. Senkevich and S.B. Desu, Applied Physics Letters 72(2), 258(1998).

[170] H.-S. Noh, Y. Choi, C.-F. Wu, P.J. Hesketh, and M.G. Allen, Transducers 03 Vol. 1, 798(2003).

[171] S. Mutlu, C. Yu, F. Svec, C.H. Mastrangelo, J.M.J. Frechet, and Y.B Gianchandani, Transducers 03 Vol 1,802(2003).

[172] G.E. Loeb, R.A. Peck, and J. Martyniuk, J. Neurosci. Methods 63(1-2), 175(1995).

[173] Pacing Clin Electrophysiol (PAB), 10(2pt1), 372(1997).

[174] A.W. Hahn, H.K. Yasuda, W.J. James, M.F. Nichols, R.K. Sadhir, A.K. Sharma, O.A. Pringle, D.H. York, E.J. Chalreson, Biomed. Sci. Instrum. 17, 109(1981).

[175] E.M. Schmidt, J.S. McIntosh, M.J. Bak, Med. Bio. Eng. Comput. 26, 96(1988).

[176] T.G. Yuen, W.F. Agnew, and L.A. Bullara, Biomaterials 8(2), 138(1987).

[177] G.E. Loeb, M.J. Bak, M. Salcman, and E.M. Schmidt, IEEE Transactions on Biomedical Engineering 24(2), 121(1977).

[178] A.B. Fontaine, K. Koelling, S.D. Passos, J. Cearlock, R. Hoffman, D.G. Spigos, J. Endovasc. Surg. 3(3), 276(1996).

Index